CW01335129

WEATHER FORECASTING HANDBOOK

TIM VASQUEZ

WEATHER GRAPHICS TECHNOLOGIES

Fifth edition
May 2002 printing

Copyright ©1994, 2002 Tim Vasquez
All rights reserved

For information about permission to reproduce selections from this book, write to Weather Graphics Technologies, P.O. Box 450211, Garland TX 75045, or to support@weathergraphics.com . No part of this publication may be reproduced, stored in a retrieval system, or transmitted by any means without the express written permission of the publisher.

ISBN 0-9706840-2-9

Printed in the United States of America

Weather Graphics Technologies
P.O. Box 450211 Garland, TX 75045
(888) 388-0070 fax (405) 329-5275
Web site: www.weathergraphics.com
support@weathergraphics.com

Contents

1 FUNDAMENTALS / 2
1.1 UNITS / 3
1.2. PHYSICAL TERMS / 4
1.3. SCALE / 8
1.4. ATMOSPHERIC STRUCTURE / 9
1.5. PRESSURE COORDINATE SYSTEM / 10
1.6. GLOBAL CIRCULATION / 11
1.7. KEYS TO METEOROLOGY / 13

2 OBSERVATION / 18
2.1. OBSERVATION NETWORK / 19
2.1. OBSERVATION CODING FORMATS / 20
2.2. OBSERVATION ELEMENTS / 22

3 TOOLS / 30
3.1. CHART ANALYSIS / 32
3.2. RADAR / 41
3.3. PROFILER / 43
3.3. SATELLITE / 45
3.4. SOUNDINGS / 47
3.5. HODOGRAPHS / 49
3.6. LIGHTNING DETECTION / 54

4 PHYSICS / 58
4.1. PHASE CHANGES / 59
4.2. ADIABATIC CHANGES / 59
4.3. STABILITY / 60
4.4. ATMOSPHERIC FORCES / 64
4.5. WINDS / 65

5 FRONTS AND JETS / 70
5.1. AIR MASSES / 71
5.2. FRONTAL CONCEPTS / 72
5.3. COLD FRONT / 75
5.4. WARM FRONT / 75
5.5. QUASISTATIONARY FRONT / 77
5.6. OCCLUDED FRONT / 77
5.7. DRYLINE / 78
5.8. SEA/LAND BREEZE FRONTS / 80
5.9. JETS / 80

6 MOTION / 84
6.1. LONG WAVES / 85
6.2. SHORT WAVES / 88
6.3. DIVERGENCE/CONVERGENCE / 88
6.4. VERTICAL MOTION / 90
6.5. JET STREAK DYNAMICS / 93
6.6. THERMAL ADVECTION / 95
6.7. VORTICITY / 96

6.8. Q-VECTORS / 102
6.9. ISENTROPIC ANALYSIS / 103
6.10. CONDITIONAL SYMM. INSTABILITY / 106

7 BAROTROPIC SYSTEMS / 110
7.1. COLD CORE BAROTROPIC LOW / 111
7.2. WARM CORE BAROTROPIC LOW / 112
7.3. COLD-CORE BAROTROPIC HIGH / 112
7.4. WARM-CORE BAROTROPIC HIGH / 113

8 BAROCLINIC SYSTEMS / 120
8.1. BAROCLINIC LOW / 121
8.2. BAROCLINIC HIGH / 122

9 CONVECTIVE WEATHER / 126
9.1. THUNDERSTORM STRUCTURE / 127
9.2. THUNDERSTORM TYPES / 128
9.3. SUPERCELL / 132
9.4. WIND PROFILES / 133
9.5. PREDICTION INDICATORS / 136

10 WINTER FORECASTING / 142
10.1. PRECIPITATION TYPE / 143
10.2. HORIZONTAL ANALYSIS / 143
10.3. VERTICAL ANALYSIS / 144
10.4. DIABATIC CHANGES / 145
10.5. SNOW CHARACTERISTICS / 147

11 TROPICAL WEATHER / 150
11.1. EQUATORIAL TROUGH (ITCZ) / 151
11.2. SUBTROPICAL RIDGE / 152
11.3. TRADE WINDS / 152
11.4. SUBTROPICAL STATIONARY FRONTS / 152
11.5. EASTERLY WAVES / 153
11.6. MID-TROPOSPHERIC CYCLONES / 154
11.7. TUTT / 154
11.8. TROPICAL CYCLONES / 154

12 / NUMERICAL GUIDANCE / 160
12.1. PREDICTION PROCESS / 161
12.2. CLASSES / 162
12.3. DOMAINS / 162
12.4. CONFIGURATION / 163
12.5. MODEL TYPES / 163
12.6. LIMITATIONS OF MODELS / 166

APPENDIX / 169
ANALYSIS EXERCISES / 181
INDEX / 195

PREFACE

Ever since the 1970s when I started digging around in libraries and querying publishers for good weather books, I've found that the available titles are consistently in one of three genres: *anecdotal books* for the layperson, *dry elementary textbooks* for college students, and *complex research titles*. In spite of the widespread public availability of weather data that blossomed in the 1990s, there are still no titles that give a ground-up introduction to forecasting for advanced hobbyists, storm chasers, pilots, media forecasters, and enthusiastic professionals. The title you're now holding is what fills the void. The concept of hobbyists quietly reading barometers and listening to official forecasts is a relic of the 1950s. Today's hobbyists are highly active, awed by the always-changing weather patterns. They use all sorts of fascinating tools such as radar loops, water vapor imagery, upper air maps, profiler plots, and hand-analysis of weather data.

A meteorology degree and an understanding of advanced math will go a long way towards the mastery of forecasting, however what is most essential are three things: *1) a strong foundation in analysis; 2) frequent practice of analysis; and 3) the willingness to learn*. Storm chasers bear out this principle quite well. The vast majority don't have degrees or math knowledge, but they are expert analysts, practice their skills quite frequently, and always comb through the past day's data when they miss out on that big tornado outbreak. As a result, these individuals have tremendous insight into the workings of severe weather and often end up converging in the same remote town hours before storms form. Some of the wisest and most amazing forecasters I know of are storm chasers. There's much to learn from them.

Conversely, readers will notice that I have planted numerous caution flags around the topics of numerical modelling. There has been a remarkable and dangerous trend toward embracing model output, particularly among media forecasters, university degree programs, and inexperienced meteorologists at all levels. While models are highly useful for an assortment of tasks and goals, they are often used inappropriately, especially in fast-breaking convective situations, and are often used as a method for bypassing "frequent practice of analysis" (see above). This essentially shortcuts the most important part of forecasting: diagnosis. Part of the goal of this book is to show the value of observed data and meteorological fundamentals in mentally visualizing the atmosphere. Indeed, the automobile owner who takes time to visualize how the car works, tinkers often under the hood, and learns how to do his own maintenance often has a car in vastly better shape than the one who simply relies on various auto shops. Not only does such a person develop unparalleled intuition and wisdom about auto repairs, they also know exactly when to depend on the auto shop and where they excel.

Overall, my aim in writing the *Weather Forecasting Handbook* is that it serve as a readable, comprehensive, yet technical introduction for amateurs with a little bit of fun and humor thrown in. It is not designed to be a dry, exacting textbook or a scientifically-groundbreaking work. Therefore there isn't a summary of references, there's plenty of educational trivia, and buried in the pages is a favorite recipe of mine.

If an error or typo is discovered in this text, I ask that you do me the professional courtesy of informing me at tim@weathergraphics.com so that it can be added into our errata sheet. All corrections and suggestions -are- integrated into upcoming editions and are compiled on our official web site. I may be out in left field but I feel that the responsibility of making a book truly useful does not stop once it's in print. Be sure to visit the *Weather Forecasting Handbook* site at www.weathergraphics.com/fcstbook for downloadable supplements, errata, and additional information.

For helping to catch errors, I owe special thanks to Pete Robinson, Keith Heidorn (The Weather Doctor), Andrew Clausen, Derek Rosendahl, and Michael Vogt. I thank Matt Bunkers for supplying me with a copy of his latest storm motion work. If there's anyone I forgot, I apologize — let me know and I'll add you to the next edition's preface.

Finally, I apologize that this book isn't a slick 4-color glossy title. There's a saying in the print industry that to go full-color you pay $100,000 for the first copy and $1 for each additional copy. There's quite a bit of truth to that! Unfortunately I don't think anyone's willing to pay $95 for this book (unless I make it into a college textbook, but I'll save that rant for another time!) My goal is to make this book clear, captivating, and low-cost so that it doesn't require any of those other frills — and as informative as possible.

Enjoy!

Tim Vasquez
May 2002
Norman, Oklahoma
tim@weathergraphics.com

Prescription without diagnosis is malpractice. Both in medicine and meteorology.

-- Anonymous

Stratus and altocumulus over Prince Rupert, British Columbia. *(Tim Vasquez)*

1 FUNDAMENTALS

The ocean forms a vast ocean above us, an ocean but little explored. We crawl about on the ground like crabs on the bottom of the sea. We make our meteorological observations down on the ground, ignorant of all that is going on in the midst of that great expanse of air above our heads, where the clouds hang about, where the rain and hail are formed, where the lightning flashes have their origin.

ROBERT BADEN-POWELL
Quarterly Journal of the Royal Meteorological Society, 1907

This chapter forms a foundation for the material that will follow throughout this book. It explains how we express physical measurements, what type of measurements can be taken, and then constructs a framework of our atmosphere and its circulation.

1.1. UNITS

The international system for measuring various physical parameters is the SI (International System) of units. This expresses length in terms of meters, mass in kilograms, and temperature in degrees Kelvin.

Unit	SI units	Abbv.	Nonstandard equiv.
Distance	meter	m	3.28084 feet
Mass	kilogram	kg	2.2046 pounds
Time	second	s	—
Pressure	pascal	Pa	0.00001 bar
Energy	joule	J	10,000,000 ergs
Frequency	Hertz	Hz	1 cps (cycle per second)
Temperature	Kelvin	deg K	-273.16 deg Celsius

Furthermore, the basic units can be expressed as a factor, for example instead of writing "5,000 meters" we can use the prefix kilo- to express the value as "5 kilometers".

Multiple	Prefix	Abbv.	Weather usage
1,000,000	mega-	M-	MHz (radio frequency)
1,000	kilo-	k-	km (distance); kg (mass)
100	hecto-	h-	hPa (pressure; 1 hPa = 1 mb)
10	deka-	da-	dam (geopotential height)
1	—	—	—
0.1	deci-	d-	—
0.01	centi-	c-	cm (radar wavelength)
0.001	milli	m-	mm (droplet size)
0.000001	micro	μ-	μm (particle size)

Although it would be ideal to use all of these units exclusively in weather forecasting, custom and convention sometimes dictate the use of antiquated systems, particularly in the United States. Such exceptions will be highlighted in the next section where appropriate.

Watch those units
The Mars Polar Orbiter, which was launched in December 1998, was destroyed when it unexpectedly plunged into the Martian atmosphere and crashed. The cause? A contractor, Lockheed Martin, provided a table of navigation data in units of "pounds per second" to the Jet Propulsion Laboratory (JPL). They were expecting "grams per second".

1.2. PHYSICAL TERMS

It's important to understand the forces that act on a parcel (a tiny, imaginary cube) of air moving through the atmosphere. These forces are expressed through physical terms and equations, and are the building blocks for learning about the atmosphere. A little algebra knowledge will help you understand the relationship between some of these quantities.

1.2.1. Mass. Mass defines how much matter an object contains in terms of subatomic particles. It is functionally the same as weight when Earth's gravity is present. It is expressed in kilograms or grams. One kilogram equals 2.2 pounds.

1.2.2. Distance. Distance is the amount of space separating two points. Derivations of distance include area (two-dimensional), volume (three-dimensional), and velocity (distance covered in a given time). According to SI units, distance should always be measured in meters or kilometers. Unfortunately nonstandard units are still in widespread use, particularly in the United States. Since the vast majority of readers are in the United States, this book will use nonstandard units where appropriate according to American meteorological convention, and will try to provide metric units where possible. Failing that, this table will provide the necessary conversion factors.

DISTANCE

Unit	Equals	Also equals	Also equals (ft)
ft	m x 3.2808	—	1 ft
m	ft x 0.3048	—	3.2808 ft
km	sm x 1.609	km x 0.621	3280.8 ft
sm	km x 0.621	nm x 0.8696	5280 ft
nm	km x 0.54	sm x 1.15	6080 ft

VELOCITY

Unit	Equals	Also equals	Also equals
m/s	km/h x 0.27	kt x 0.5144	mph x 0.4473
km/h	m/s x 3.6	kt x 1.85	mph x 1.609
kt	m/s x 1.944	km/h x 0.54	mph x 0.87
mph	m/s x 2.2356	km/h x 0.621	kt x 1.15

1.2.3. Pressure. Pressure is simply force per unit area. It equals $P = F/A$, where F is force (N) and A is area (m^2). The result is N/m^2, which is a Pascal (Pa). Pressure is usually expressed in millibars (mb) or hectopascals (hPa), both of which are exactly the same. The weight of the atmosphere above a unit area decreases with increasing height, therefore the pressure

---

Metric system in America
By 1968, most of the world had gone metric and the Department of Commerce began studying the feasibility of a similar move in the United States to help expand its technological progress. In 1975 Congress passed the Metric Conversion Act, establishing the U.S. Metric Board to help promote the new system to American citizens and industry. However in doing so it failed to adopt the Department of Commerce recommendation to set a completion timetable. Metric fever slowly plateaued in 1979 and began a gradual decline as the general public lost interest. Impotent from birth, the U.S. Metric Board reported in 1982 that it had done all it could and had no authority to force further changes. It was disbanded. The idea was resurrected in 1988 under the Omnibus Trade and Competitiveness Act which recommended the metric system in trade and required it in all government activities by 1993.

decreases with height. Pressure may also be measured in terms of the height of a column of mercury in an evacuated tube, and in such a case is expressed in millimeters (mm) or inches of mercury (in Hg). One inch of mercury equals 33.8636 mb.

1.2.4. Force. Force is an influence that tends to produce a change in movement or shape. It equals **F = m · a**, where m is mass (kg) and a is acceleration (m/s^2). The result is force (kg·m/s^2). This also equals a newton (N). Atmospheric pressure is a measurement of force.

1.2.5. Density. In meteorology, density describes the mass of a specific volume of air. It is expressed by the formula $\rho = m / v$, where m is mass (kg) and v equals volume (m^3). The result is density (kg/m^3). The symbol for density is ρ, the Greek letter rho. By rearranging the basic ideal gas equation P=ρrT, where P is pressure, r is constant, and T is temperature, we get the simplified equation ρ=P/(rT). Simple algebra shows that if we hold P and r constant and raise T, then ρ will decrease. This illustrates why cold air is denser than warm air.

1.2.6. Temperature. This is the degree of hotness or coldness measured with a thermometer using some definite scale. Three units of temperature are used in meteorology, described below. There is also the Rankine scale, where 0 °R is absolute zero and 491.67 °R is the melting point of water, and the antiquated Réaumur scale, with 0 °r and 80 °r at the melting and boiling point of water, respectively, but neither is used in meteorology. Note that when a difference or change in temperature is indicated, the correct terminology is "[unit] degrees" rather than "degrees [unit]", such as, "The temperature fell by 24 Fahrenheit degrees (by 24 F°). This book will adhere to this convention.

* *Fahrenheit*. The melting point is 32 deg F and the boiling point is 212 deg F. To convert to Fahrenheit, use °F = (1.8×°C) +32. The Fahrenheit scale is used only in the United States for surface charts and public temperature forecasts. The scale was developed in the 18th century so that 0 °F would be the coldest possible temperature in an ice-salt mixture and 100 °F would be the human body temperature.

* *Celsius*. The melting point is 0 °C and the boiling point is 100 °C. To convert to Celsius, use °C = (°F - 32)×0.555, or °C=°K-273.16. The Celsius scale is used on all charts, except for surface charts in the United States (Fahrenheit) and where Kelvin units are required (see below).

Millibars vs. hectopascals
It is a mystery to me what was wrong with millibars, a perfectly good metric unit. It's not as if we need a nomogram or a calculator in going from mb to hPa, after all! . . . It's not at all obvious that our science is being advanced in any way by converting from mb to hPa.
CHUCK DOSWELL,
"Pet Peeves of Chuck Doswell," 2000

HUMOR BREAK
The Laws of Thermodynamics are the building blocks for learning about energy exchange between a system and its surroundings.

First Law: The energy of the universe is constant; energy cannot be created or destroyed, but only changed from one form to another.
Second Law: In all energy transformations, there is an irreduceable fraction which is converted to a form which can no longer do work.
Third Law: The entropy of a perfect crystalline substance at the temperature of absolute zero is zero.

An unknown physicist once paraphrased these laws in less abstract terms:

First Law: You can't win. You can only break even.
Second Law: You can only break even at absolute zero.
Third Law: You can never reach absolute zero.

Wet bulb

Humidity is generally measured by determining the rate of evaporation of water into the air. This is generally accomplished by the use of a psychrometer, an instrument consisting of two thermometers, the bulb of one being surrounded by a moistened cloth from which water is free to evporate and the other being freely exposed to the air. Due to evaporation, the temperature indicated by the thermometer with the "wet-bulb" will be somewhat lower than the other, the more rapid the evaporation the greater the difference between the two thermometers. When the air is saturated the two thermometers will indicate the same temperature, while if it is very dry the wet-bulb thermometer may indicate a temperature many degrees lower than the other. By the use of suitable tables this difference in temperature may be used in calculating the humidity.

GEORGE TAYLOR
"Aeronautical Meteorology", 1938

Kelvin. Kelvin is simply the Celsius scale offset so that a temperature of 0 °K equals absolute zero rather than the melting point of water. It is a base unit in the SI measurement system. In degrees Kelvin, the melting point is 273 °K and the boiling point is 373 °K. To convert a Celsius reading to Kelvin, use °K=°C+273. By convention the Kelvin scale is used only in advanced meteorological products that use potential temperature (to be discussed shortly).

1.2.7. Dewpoint temperature (T_d), in degrees (computation formula not presented here due to its complexity). Assuming the pressure and moisture remains constant, this is the temperature at which saturation will occur if the air is cooled. When this point is reached, water droplets will usually begin condensing out of the air, forming clouds and possibly precipitation. Dewpoint is dependent only on the amount of water vapor in the air, so it makes a good indicator of moisture. It cannot be directly measured, but it can be observed if the air is cooled or lifted. Dewpoint is reported by nearly all weather stations.

1.2.8. Dewpoint depression, in degrees. The equation is simply $T_{dd} = T - T_d$. Indicates the number of degrees the dewpoint temperature is below the actual temperature. It illustrates how close the air is to saturation. Small values indicate the air is close to saturation, while high values indicate the air has little moisture.

1.2.9. Wet bulb temperature (T_w), in degrees. Indicates the temperature to which air is cooled by evaporation until saturation occurs. It can be directly measured using a wet-bulb thermometer, aspirating a wet wick and reading the lowest measured temperature. The wet bulb temperature always lies somewhere between the temperature and the dewpoint (usually a little closer to the dewpoint).

1.2.10. Mixing ratio (w), in g/kg (no formula presented here due to its complexity). Represents the ratio of the mass of water vapor to the mass of dry air in a given volume. Mixing ratio is an excellent way to measure the exact amount of moisture in the air . Mixing ratio is independent of temperature but inversely proportional to pressure.

PHYSICS • 7

1.2.11. Saturation mixing ratio (w_s), in g/kg (formula not presented here due to its complexity). This quantity represents the maximum possible mixing ratio for a given parcel of air. It is directly proportional to temperature. The ratio of w to w_s equals the relative humidity.

1.2.12. Virtual temperature (T_v), in degrees. This is temperature taking density into account. It equals the temperature that air without any water vapor at all would have if it were the same density as a given parcel. It equals **$T_v = T + (w/6)$**, where T is the ambient temperature in Celsius and w is the mixing ratio in g/kg. By increasing the moisture (dewpoint), T_v will be as much as several degrees higher than the ambient air temperature (moist air is less dense than dry air, and hot air is less dense than cold air).

1.2.13. Relative humidity (RH), in percent. Provides a ratio of the amount of water vapor to the saturation value. Calculated as RH = w / w_s ·100.

1.2.14. Potential temperature (θ or theta), in degrees. Indicates the temperature a parcel would have if it were brought dry adiabatically to 1000 mb. This provides an accurate way of comparing the temperatures of parcels at two different levels.

1.2.15. Equation of state. The equation of state, also known as the ideal gas law, is useful for understanding the relationship between pressure, temperature, and density. It equals **P = ρRT**, where P is pressure, ρ is the density (m/V), R is the gas constant for dry air (2.87), and T is the temperature in Kelvin. A simple way to use it is to algebraically hold two of the quantities constant and observe which way a change in one quantity affects another. This yields the following relationships:

* *Constant temperature.* At a constant temperature, pressure and density are directly proportional. So if it's 50 deg in Dallas and 50 deg on Pike's Peak, the density will be lower on Pike's Peak where pressure is lower.

* *Constant pressure.* At a constant pressure, density and temperature are inversely proportional. So if New York and Miami have the same pressure but New York is colder, the density in New York will be higher.

* *Constant density.* At a constant density, pressure and temperature are directly proportional. So if a given volume of air

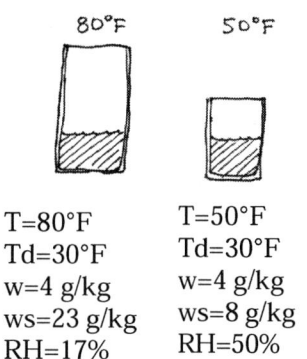

T=80°F
Td=30°F
w=4 g/kg
ws=23 g/kg
RH=17%

T=50°F
Td=30°F
w=4 g/kg
ws=8 g/kg
RH=50%

Figure 1-1. Simplistic *analogy* of how moisture variables change when the temperature changes. In this example, a parcel of air is represented by a glass of water. The parcel contains 4 g/kg of water vapor. When we cool the air, the so-called "holding capacity" of the glass shrinks. The saturation mixing ratio decreases and the relative humidity goes up. Although this is an excellent analogy that allows beginning forecasters to understand how moisture values change, it is important to remember that air does not physically have a "holding capacity" like a glass or sponge.

descends from 30,000 ft to 10,000 ft, pressure will naturally increase, causing the temperature to rise.

1.2.16. Hydrostatic equation. Since a vacuum exists above the earth, the atmosphere has an inclination to rush out into space to fill the void. It doesn't, though, due to the pull of gravity. This balance is called hydrostatic equilibrium, and is illustrated by the hydrostatic equation: $\Delta P/\Delta Z = -\rho g$. The left term shows the vertical pressure gradient (change in pressure per change of vertical distance) and the right term represents density multiplied by gravity.

1.2.17. Hypsometric equation. This equation indicates the thickness of a layer (bounded by two pressure surfaces) in weather forecasting. The conceptual equation is $\Delta Z = k T_v$, where ΔZ is thickness in meters, k is a constant unique to each layer, and T_v is the mean virtual temperature of the layer. This shows that thickness (difference in distance) of a layer is directly proportional to its mean virtual temperature, and is proportional to moisture. As mixing ratio increases, virtual temperature increases and thus thickness increases.

1.3. SCALE

Weather systems and circulations are described in terms of their scale. Rather than speaking of a "low pressure area that is 400 miles wide", forecasters refer to the feature as a mesoscale low. These categories are not meant to divide weather systems into a spectrum of discrete types, but rather to serve as a framework for understanding and describing processes that take place in the atmosphere. It should be noted that there is not yet an accepted standard for scales; they tend to vary in various sources.

1.3.1. Planetary scale. Scales on the order of 10,000 km or greater. This describes the general circulation and east-west extent of jet streams.

1.3.2. Synoptic scale. Scales on the order of 1,000 to 10,000 km. It encompasses baroclinic waves, large frontal systems, long waves, etc.

1.3.3. Subsynoptic scale. Scales on the order of 100 to 1,000 km. It encompasses squall lines, tropical cyclones, short waves, etc. This scale is sometimes loosely referred to as "mesoscale" (see below).

Hydrostatic equlibrium
Hydrostatic equilibrium not only determines the structure of the earth's atmosphere with respect to gravity. It is also used by astrophysicists to describe the balance of a star's energy with respect to gravity. If the energy output is the same over a given period, the star will neither expand nor contract, and it is said to be stable.

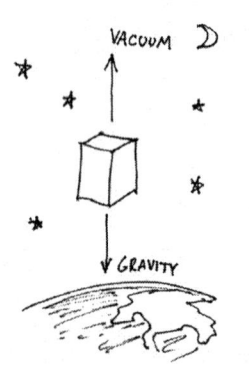

Figure 1-2. Hydrostatic balance. The tendency for a parcel of air to escape into the vacuum of outer space is balanced by the attraction of gravity. This balance of forces is known as hydrostatic balance.

PHYSICS • 9

1.3.4. Mesoscale. Scales on the order of 10 to 100 km (sometimes defined as 10 to 1,000 km). It includes thunderstorms, cloud clusters, thunderstorms, coastal fronts, drylines, orographic effects, and so forth.

1.3.5. Microscale. Scales on the order of 10 km or less. This includes turbulence, small clouds, tornadoes, dust devils, mountain waves, etc.

1.4. ATMOSPHERIC STRUCTURE

Although the vast majority of weather occurs in the lowest layer of the atmosphere, it is essential to have a brief understanding of higher layers.

1.4.1. Troposphere. The troposphere is the lowest layer of our atmosphere, extending from the surface to a height of about 4 miles (7 km) in polar regions to 11 miles (17 km) in tropical regions. The vast majority of weather occurs in the troposphere. The top of the troposphere is called the tropopause.

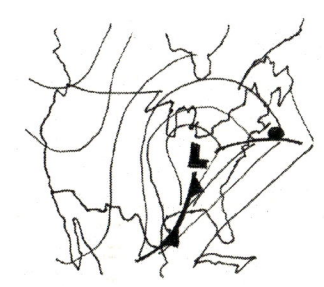

Figure 1-3. Synoptic scale. A large low pressure area is an example of a synoptic scale circulation. Since observations at the synoptic scale are readily available, it is easy to forecast synoptic scale circulations.

1.4.2. Stratosphere. The stratosphere is a layer above the troposphere where the temperature remains constant or increases with height. It extends from the tropopause up to the stratopause at a height of about 30 miles (50 km). The temperature increase with height is due to absorption of solar heat by a layer of ozone within the stratosphere. Although the stratosphere is generally not important to forecasting, pockets of warm air in the lower stratosphere (called "warm sinks") are often associated with regions of strong upward motion in the troposphere. Nacreous clouds, also known as "mother of pearl" clouds, occasionally occur in polar regions at a height of 15 to 18 miles (25 to 30 km), however they are currently of no concern to operational meteorologists.

1.4.3. Mesosphere. This is a layer above the stratosphere, extending between the stratopause and the mesopause, at a height between 30 and 50 miles (50 and 80 km). In the mesosphere the temperature decreases with height. It is not significant in day-to-day forecasting. The top of the mesosphere is called the mesopause. Near the mesopause, noctilucent clouds are occasionally observed in polar regions, however no connection has been found with tropospheric weather. Noctilucent clouds look like pale silvery cirrus during the day, and by night they glow in the sun's light for many hours just

Figure 1-4. Microscale. A dust devil is an example of a microscale circulation. Since a network of microscale observations does not exist, it is impossible to accurately predict microscale circulations. Imagine even trying to deploy a network of sensors in this cow pasture and rapidly assimilate the data.

The first high altitude journey

In the year 1850, M. M. Barral and Bixio conceived the project of ascending to a height of 30,000 or 40,000 feet, for the purpose of investigating certain atmospheric phenomena still imperfectly understood . . . Nothing daunted by the ill-success of their first expedition, and eager to obtain a better result from a second trial, Barral and Bixio determined to ascend again without delay. On July 27, 1850, the filling of the balloon was commenced early in the morning. It proved to be a long operation, occupying till nearly two o'clock. The sky became overcast, and it was past four when they left the earth. They soon entered a cloud at 7,000 or 8,000 feet, which proved to be fully 15,000 feet in thickness. They never, however reached its highest point, for when at 4 h. 50 m. the height of 23,000 ft was attained, they began to descend, owing to a rent which was then found in the balloon. After vainly attempting to check this involuntary descent, they reached the earth at 5h 3s. The most unexpected result observed in this ascent was the extraordinary decrease of temperature . . . When near the highest point which they attained, their thermometer sank to -38.2°F. The clothes of the observers were covered with fine needles of ice. 'This discovery,' says Arago, in his Report to the Academy, 'explains how these minute crystals may become the nuclei of large hailstones, for they may condense around them the aqueous vapor contained in that portion of the atmosphere where they exist.' . . . The great extent of so cold a cloud explains very satisfactorily the sudden changes of temperature which occur in our climates.

DR. G. HARTWIG
"The Aerial World," 1886

---

before or after twilight, bewildering tourists and visitors to the north.

1.4.4. Thermosphere. Existing above the mesosphere, the thermosphere (also known as the ionosphere) is a layer in which temperature increases with height. This layer is not significant in day-to-day forecasting. The temperature increase is due to absorption of solar heat by hydrogen ions. The thermosphere is important in AM and shortwave radio broadcasting because it contains layers that are reflective to radio waves. The thermosphere is also the playing grounds of the polar aurora, which occurs at a height of 125 to 300 miles (200 to 500 km). The aurora, caused by corpuscular solar radiation streaming into the rarified gases of the outer atmosphere, appears as dancing, rippling glows of all different colors and is often a nightly treat for residents of the northern lands.

1.5. PRESSURE COORDINATE SYSTEM

When you hear "that plane's at 35,000 feet" or "the height of Elvis' UFO was 700 feet", you are hearing a reference to a height coordinate system. Meteorologists tend to use a pressure coordinate system, however; and might say "that plane is at the 200 millibar level" or "the height of Elvis' UFO was near the 950 mb level". Pressure coordinate systems are used because meteorological processes are dominated by physical interactions based on temperature, pressure, and density (height is a man-made property that requires measurement). This simplifies analysis tasks and calculations, and it continues in standard use today. To better explain this system, the 500 mb level, for example, represents the height to which a barometer would have to ascend to read 500 mb. This level, when visualized over a broad horizontal area, is referred to as a *constant pressure surface*. The actual height of a constant pressure surface varies according to the atmospheric density below that particular point on the surface plane.

1.5.1. Standard height equivalents. To properly put pressure levels into perspective, it is necessary to approximate their height, and vice versa. Here are the standard levels and processes found at each level, assuming a standard atmosphere. Proficient forecasters have these equivalents memorized.

* 1000 mb (300 ft (100 m) MSL). Reflection of surface conditions.

* 850 mb (5000 ft (1.5 km) MSL). Storm forecasting and evaluating low-level winds with friction effects minimized.
* 700 mb (10,000 ft (3 km) MSL). Short wave disturbances, mid clouds.
* 500 mb (18,000 ft (6 km) MSL). Short wave disturbances, mid clouds
* 300 mb (30,000 ft (9 km) MSL). Long wave disturbances, wintertime analysis of the jet stream.
* 200 mb (39,000 ft (12 km) MSL). Long wave disturbances, summertime analysis of the jet stream.
* 100 mb (54,000 ft (17 km) MSL). Mostly used in the tropics to analyze the upper troposphere

1.5.2. Variation of height. A two-dimensional surface of equal pressure is referred to as a pressure surface. The height of this surface is not constant due to density variations in the atmosphere. Forecasters map out the height of pressure surfaces using lines of equal height (contours). Areas of high height equate to areas of high pressure at that level (the surface is bumped upward from below by higher values of pressure). Likewise, "low height" loosely equates to "low pressure".

1.6. GLOBAL CIRCULATION

The global circulation is a model that attempts to describe the planetary motion of air. In an simplified atmosphere, these patterns would dominate the weather, but it is now known that two of the cells that make up the three-cell global circulation model presented here are usually dominated by complex physical processes and are rarely a factor in everyday forecasting.

1.6.1. Hadley cell. This concept, developed in 1735 by meteorologist George Hadley, describes the heat-driven circulation that originates at the equator. Solar heating heats the atmosphere directly on the equator. This air rises and flows toward the poles. It was originally thought that this air went all the way to the poles, forming the overall global circulation, but it was later demonstrated that the Coriolis force (to be discussed shortly) begins deflecting the air (rightward in NH, the northern hemisphere), where it accumulates in the subtropical latitudes. This forces the air to sink toward the surface. The result is a band of sinking air at approximately 30 degrees latitude, generally called the subtropical high. Air sinks to the surface then flows back toward the equator. It is deflected rightward (NH), resulting in a flow of prevailing winds moving from northeast to southwest (NH). This wind is known as the trade

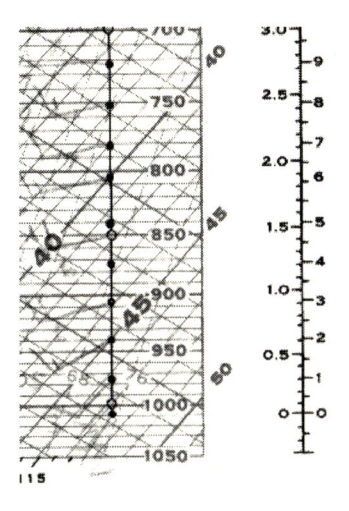

Figure 1-5. The scale of the SKEW-T sounding chart (to be discussed soon) shows the relationship of pressure and height in a standard atmosphere. The numbers reading 600 through 1050 on the vertical center scale are pressures in millibars. The numbers reading 0 through 4.0 on the inner right vertical scale are heights in kilometers. The numbers reading 0 through 14 on the outer right vertical scale are heights in thousands of feet. So the 850 mb level is at a height of about 1.5 km, or 5000 ft. Note how pressure drops with increasing height.

Pop quiz!
QUESTION: You are on an MD-80 flight, just now cruising after leaving Los Angeles. Your pocket altimeter read 0 ft at the gate but now reads 6,000 ft. You remember that your ears popped during the climbout. What percentage of the air that was in the plane is gone now? You are reading the Weather Forecasting Handbook and have the above diagram handy.
ANSWER: The diagram shows that the pressure at the gate was about 1000 mb and at 6,000 ft is about 800 mb. Therefore about 20% of the air has been bled from the cabin.

Hadley's wind theory

"The same principle, as necessarily extends to the production of the west trade winds without the tropics. The air rarefied by the heat of the sun about the equatorial parts, being removed to make room for the air from the cooler parts, must rise upwards from the earth, and as it is a fluid, will then spread itself abroad over the other air, and so its motion in the upper region must be to the north and south from the Equator. Being got up at a distance from the surface of the earth, it will soon lose a great part of its heat, and thereby acquire density and gravity sufficient to make it approach its surface again, which may be supposed to be by the time it is arrived at those parts beyond the tropics where the westerly winds are found. Being supposed at first to have the velocity of the surface of the earth at the Equator, it will have a greater velocity than the parts it now arrives at, and thereby become a westerly wind, with strength proportional to the difference of velocity, which in several revolutions will be reduced to a certain degree, as is said before, of the easterly winds, at the Equator. And thus the air will continue to circulate, and gain and lose velocity by turns from the surface of the earth or sea, as it approaches to, or recedes from the equator."

GEORGE HADLEY
"Concerning the Cause of the General Trade Winds", 1735

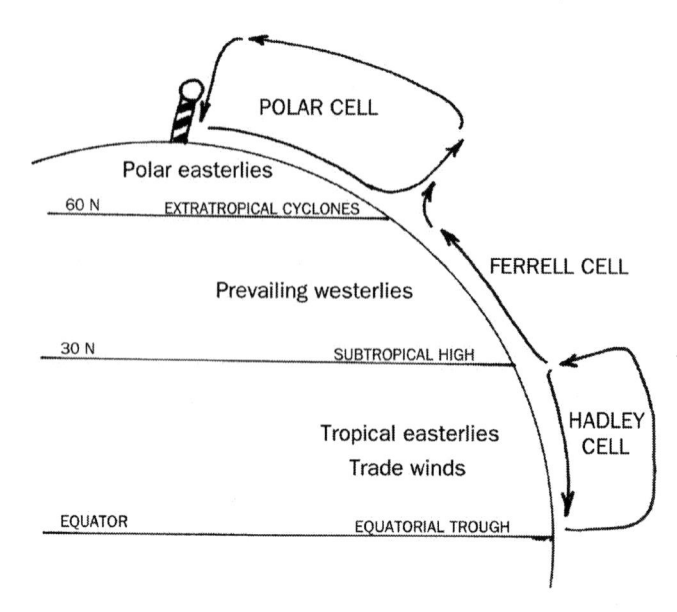

Figure 1-6. Simple model of the Earth's global circulation. The earth is divided into three main circulation cells and three main bands of wind.

winds or the tropical easterlies. It explains why winds usually blow from the east in tropical locations.

1.6.2. Ferrel cell. This is a concept that describes the circulation that exists between 30 deg latitude and approximately 60 deg latitude. It was proposed in 1856 by American mathematician William Ferrel (1817-1891). Surface air originating from the sinking region of the Hadley cell flows poleward, then begins deflecting (rightward in the NH) at about 60 degrees latitude due to the Coriolis force. The result is a band of prevailing winds that flow from southwest to northeast (NH) at the surface in mid-latitudes. This wind is known as the prevailing westerlies. Ideally the air would rise at 60 deg latitude and return equatorward, deflecting right (NH) to form an easterly upper-level wind, but this does not actually occur. The Ferrel cell is generally an oversimplification, but it does have importance in describing the "prevailing westerlies".

1.6.3. Polar cell. This cell describes the conceptual circulation that exists between the north pole, where excess radiation of heat into space continually creates a cold surface air mass, and 60 degrees latitude. Cold air from the arctic regions flows south and is deflected by the Coriolis force at about 60 degrees latitude, where it accumulates and is forced to rise. It rises, then diverges and flows back northward toward the poles. In reality this cell is also an oversimplification since flow in the high latitudes is dominated by lows, highs, and atmospheric waves. However the polar cell does have importance in describing the concept of "polar easterlies".

1.7. KEYS TO METEOROLOGY

The perfect segue into the remainder of this book is Stanley D. Gedzelman's keys to meteorology as published in *Science and Wonders of the Atmosphere* (1980). This book is long out of print but the keys it presents are more important than ever to understanding meteorology. Read these keys over and over until they make sense, and you will be set for tackling the upcoming chapters!

1.7.1. The sun's heating varies over the earth and with the seasons. Areas directly under the sun (the tropics) receive maximum heating, while areas of Earth that are sloped away from the sun (polar regions) receive less heating per unit area. This temperature gradient is the fuel that drives the engine of the atmosphere. Without this unequal heating, there would be no weather.

1.7.2. The differences of air temperature over the Earth cause the winds. Warm air rises and because of the structure of the atmosphere is ultimately forced to spread outward (diverge). This removes mass above the surface, directly resulting in pressure falls at the surface. Other processes aloft indirectly caused by heating can accomplish the same thing. The result is a partial vacuum condition at the surface that causes surface air to converge inward to fill the void. This process occurs all the time, in all scales and places.

1.7.3. The rotation of the Earth destroys this simple wind pattern, twisting the winds and producing great wind spirals that are known as high and low pressure areas. The Earth's rotation produces an effect called the Coriolis force, which bends air to the right in the northern hemisphere and to the left in the southern hemisphere. Near the Equator it has little or no effect. The Coriolis force significantly reduces the ability of air to move around and "fill the void", which prolongs and complicates the life cycle and structure of high and low pressure areas.

Tidally locked worlds
Science-fiction authors and fans often speculate what the Earth would be like if it stopped rotating. Would the sunward side become an oven with a frozen night side?
Jason C. Goodman, an experienced climatology graduate student, speculated that the loss of the Earth's rotation would vastly improve the efficiency of the atmosphere as an engine to dispel temperature differences. In fact, the temperature difference between the night and day side might be as little as 10 degrees C. However between these zones the atmospheric engine would rage in full force, with a 500 mph windstorm transporting cold air from the night side to the day side, and aloft a strong jet stream transporting warm air to the night side. The livable regions would be the center of the day side, where light winds, torrential rain, and warm weather would predominate, and on the center of the night side, where light winds, clear skies, and cool weather would be found.
Several physicists tended to agree with the assessment, figuring a night temperature near freezing and a dayside temperature just above 130 deg F, but with much lighter winds between the two zones.
Still interested? Check out "Simulations of the Atmospheres of Synchronously Rotating Terrestrial Planets Orbiting M Dwarfs: Conditions for Atmospheric Collapse and the Implications for Habitability" by Joshi, Haberle, and Reynolds (Icarus V129, pp 450-465, 1997).

14 • WEATHER FORECASTING HANDBOOK

1.7.4. Since cool air can "hold" less water vapor than warm air, rain and other forms of precipitation are caused by cooling the air. The ability of air to retain water vapor decreases if it is cooled. Eventually it will reach a temperature at which no more water vapor can be held. Water then condenses out of the air in the form of visible droplets: clouds and possibly precipitation.

1.7.5. Pressure in the atmosphere decreases with increasing height. Anyone who has had trouble breathing on top of a mountain or is scared of being sucked out of an airliner at cruise altitude knows this fact. Gravity keeps air molecules packed close to the Earth. Pressure, as measured by a barometer, is determined by the "weight" of molecules above a given point. As you rise upward, there are less molecules above you. Therefore the pressure decreases.

1.7.6. Decreasing the air pressure causes the temperature to drop. This is directly observed by the way that it gets cooler as you ascend to the top of a mountain. The temperature outside of a jet airliner is typically around minus 40. In outer space, where there is no pressure, the temperature is the frigid value of absolute zero. Note that this rule does not imply that low pressure areas are cold or high pressure areas are warm; there are too many other factors at work to be able to make this assumption. But it is a very accurate rule when looking at smaller-scale phenomena and vertical motion.

1.7.7. Clouds and precipitation are caused by rising air, while clear weather is caused by sinking air.
When air is forced to rise, pressure drops as it ascends (see key #5) and this causes its temperature to drop (see key #6 above). This can ultimately cause water droplets to condense out of the air (see key #4). The result is clouds and possibly precipitation. Therefore rain and bad weather is typically associated with upward motion.

1.7.8. Rising air in low pressure areas causes clouds and precipitation, while sinking air in high pressure areas causes clear weather. Low pressure areas attract air from all around, causing convergence at the surface and forcing it to rise (if you put a bunch of M & M's on a table and push them together, most of them will rise upward) Likewise in high pressure areas air spreads away, causing downward motion. Downward motion always results in warming, drying of the air, and fair weather. Of course this key isn't always guaranteed. There are often low pressure areas that have good weather and high pressure areas that have storms and bad weather. But the above keys demonstrate that there is an *enhanced* possibility of bad weather

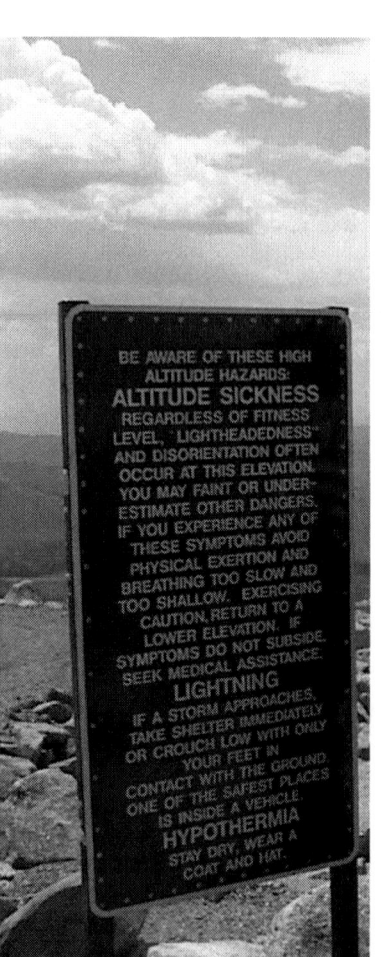

Figure 1-7. Low pressure at high altitudes is demonstrated by this sign near the summit of Mount Evans, Colorado (elevation 14,264 ft / 4348 m). It warns visitors of the dangers of high altitude sickness. The typical station pressure at this height is only 580 mb compared to the usual 1010 mb at sea level, which means almost half of the air most humans are accustomed to is gone!

in low pressure areas, and an enhanced possibility of good weather in high pressure areas. Forecasting is the fine art of figuring out what the result will be.

REVIEW QUESTIONS

1. Thunderstorms typically produce large masses of cold air known as outflow. Is outflow air denser or less dense than the surrounding air?

2. Define 300 deg Kelvin in units of Celsius and Fahrenheit.

3. An air mass is coming across the Appalachians but you have no idea if it is colder or warmer than the air mass currently at your station in Delaware. What type of temperature would help you compare the two air masses: temperature, potential temperature, or virtual temperature?

4. You are climbing Mount Everest. The air is dry but is getting progressively cooler. Finally you climb into the base of a large cumulus cloud where the air is saturated. You stop and look at your pocket thermometer, which reads 42 deg F. What is the dewpoint temperature?

5. You have analyzed a low pressure area on a surface chart. You measure the width of the circulation and find it to be 1600 km. What type of scale is this system?

6. You are analyzing an air mass. As you run your finger on the map from western Canada to eastern Canada, you note that the 500 mb heights decrease, and since surface pressures are relatively equal across Canada this tells you that the thickness of the air below 500 mb decreases from west to east. What kind of temperatures would you expect to find in eastern Canada?

7. The wild media interest in balloon expeditions continues: this time it's you taking a hot-air balloon to a height of 100,000 feet. You have an altimeter, a barometer, and a thermometer on board. What is the best way to determine when you've entered the stratosphere?

8. After giving up your meteorological career and going back to the 18th century in a time machine, you are sailing with Captain Bligh and crew in the South Pacific. It's January and you're near latitude 20 deg S. The winds have been dead for

Vertical speeds
Typical time it takes to go 1 meter (3.2 feet) up or down

UPWARD
Severe storm updraft 0.01 sec
Cumulus updraft 1 sec
Air mass on rainy day 20 sec

DOWNWARD
Air mass on fair day 3 min
Cloud droplets 1.5 min
Large cloud droplets 4 sec
Rain drops 0.25 sec
Golfball-sized hail 0.04 sec

Weight of a cloud
A cloud's weight depends on the distribution of water and ice. Typical densities range from about 0.3 to 0.7 g/m^3 in a benign weather cloud to 10 g/m^3 in a typical thunderstorm. Assuming a cumulus cloud has a density of 0.5 g/m^3 and has a size of 1 km^3 (1 billion cubic meters), we can calculate that it has a weight of 500,000 kg, or about 550 tons. This is a little heaver than the weight of a fully loaded Boeing 747.

days. The crew speaks of the dreaded doldrums, and there hasn't been a cloud in sight for a week. You and the crew are planning a mutiny, but the plan requires that you get as quickly as possible to the safe haven of Fiji to the <u>north</u>. When the winds pick up, will they blow in a favorable direction? Should the mutiny proceed?

9. You ascend in a balloon to the 500 mb level with a barometer. What will pressure will the instrument read in mb?

10. When there is a structural failure of an airliner inflight and the plane depressurizes, it is common for the cabin to fill up with fog. How can this be explained?

Barometer, National Weather Service, Albuquerque, New Mexico. *(Tim Vasquez)*

2 OBSERVATION

In my first endeavors to investigate storms in my own way, I felt the want of assistance in other parts of the country, and conceived the idea of inducing gentlemen of leisure and scientific inclinations to make voluntary observations whenever a storm passed. I desired to have them in telegraphic communication, so that when a region of low barometer formed in the south notice should at once be given to observers in the north, so that the clouds forming in advance and other phenomena should be noted.

WILLIAM BLASIUS,
"Storms: Their Nature, Classification and Laws," 1875

A study of the science of weath[er is] sometimes overlooked by for[ecasters] eager to jump in and use th[e data;] weather observation appears to be [the] groundwork that's best left to elect[ronics and] underpaid observers. However th[e forecaster must be] planted deeply in the craft of we[ather observation;] no way a forecaster can develop s[kill without an] intimate understanding of the p[rocess.]

First, weather observation [helps the forecaster] understand how information ab[out the weather] is being collected on a daily basi[s. Knowing what] instruments are being used to collect the data helps [the forecaster] question erroneous data and visualize the processes that might be occurring at the station.

2.1. OBSERVATION NETWORK

Weather observations are taken daily every 1 to 6 hours at about 5,000 stations around the world. Because of the close kinship of meteorology with aviation, it comes as no surprise that a substantial number of weather stations are located at airports and are managed by aviation authorities. This is especially true in the world's wealthier countries, including the United States, Canada, and Europe. In other countries, local weather facilities, operated by the country's weather agency, are responsible for weather observations.

The data is fed via radio, phone, or computer into a weather network that has a proud history. Its roots go back to 1951, when the World Meteorological Organization was founded as a United Nations agency. The rewards came within a few years as member nations rapidly agreed on a common set of reporting standards and established an international telecommunications network. It allowed the free flow of weather data via teletype. Feeding the network was a system of aviation teletype circuits to obtain reports from airports worldwide. The U.S. Air Force and U.S. Navy also played an integral part in the process, collecting significant amounts of data from third-world countries through its global radio intercept network and sharing this with the WMO network. At no point in international history has data been shared so freely, even across the former Iron Curtain. There has been little change over the past few decades, aside from major communications upgrades and increasing reliance on satellite delivery. The World Meteorological Organization still continues to set standard international formats for the network.

20 • WEATHER FORECASTING HANDBOOK

METAR FORMAT COVERAGE

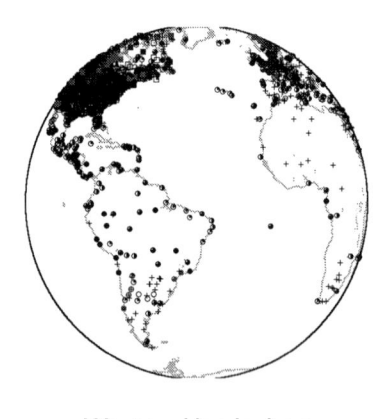

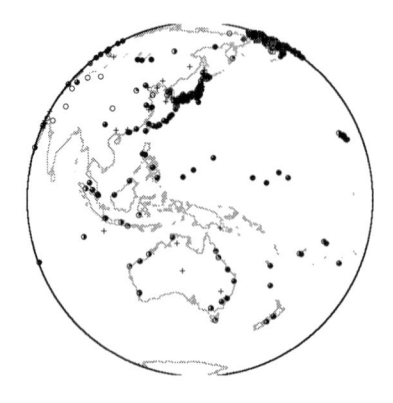

Western Hemisphere   Eastern Hemisphere

SYNOPTIC FORMAT COVERAGE

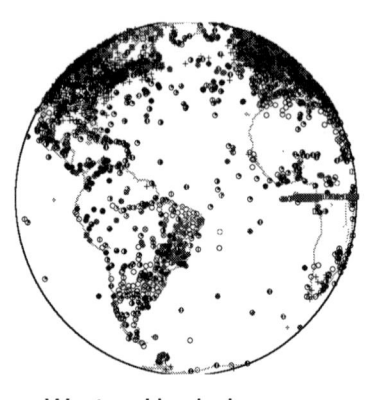

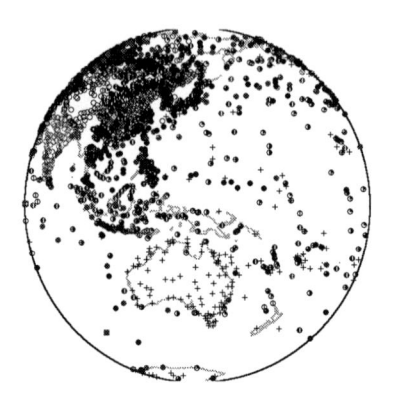

Western Hemisphere   Eastern Hemisphere

Figure 2-1. Surface data availability for October 30, 2000 at 0000 GMT. It's easy to see that METAR coverage is sporadic worldwide except in North America, Europe, and Japan. However coverage with synoptic data is fairly uniform worldwide, though not as dense as METAR in North America. Maps generated by Digital Atmosphere (www.weathergraphics.com).

OBSERVATION • 21

2.2. OBSERVATION CODING FORMATS

Three main coding formats for surface weather observations are usually encountered in meteorology. They are described briefly here.

2.2.1. METAR observations. METAR is the primary format for weather data distribution in North America, though it is used at larger airports in other regions. The format is rather readable and is designed mostly for the aviation sector. METAR observations are usually taken *every hour*. See the appendix for more information. An example:
KODX 161448Z 06005KT 15SM SCT200 28/22 A3019;

2.2.2. Synoptic observations (SYNOP). This format is used worldwide, and comprises the primary method for weather data distribution outside of North America. The format, designed mainly for meteorologists, consists of blocks of numerical data. The observations are usually taken every 6 or 12 hours (sometimes every three). An example:
17601 32575 50000 10112 20084 40197 52004 81250

2.2.3. Surface aviation observation (SAO) format. SAO format was the primary format used throughout North America before July 1996. It is no longer used except by a handful of automated stations in Canada. The format, designed for rapid teletype transmission, combined readability and brevity, and specified temperature and dewpoint in Fahrenheit. An example:
FTW SA 1650 40 BKN 7RW- 146/73/62/1518G25/990=

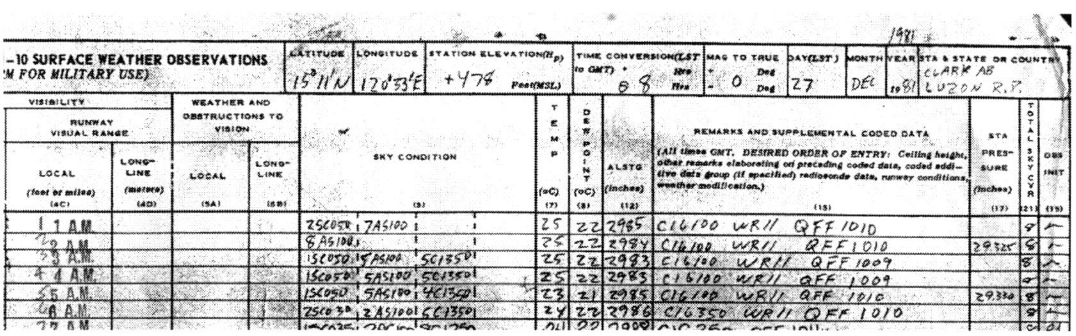

Figure 2-2. Weather observations in the United States are documented on Form 10's, like these shown here, then transmitted via government data networks. Automated observations are usually disseminated directly into the network without human intervention.

2.3. OBSERVATION ELEMENTS.

2.3.1. Temperature. Temperature is measured in a sheltered location, usually on a grassy area away from sources of unusual heating and cooling. Usually temperature is measured at a height of 1.5 m (about 5 feet). On cold, calm mornings, cold air tends to accumulate in a layer below this height, which explains why a 34 deg F (1 deg C) reading might be measured on a morning when frost is all over the ground.

2.3.2. Dewpoint. This value indicates the amount of moisture at the station. For decades it has been determined using a wet-bulb thermometer (using a wetted wick to measure evaporational cooling). However most advanced stations now use high-tech sensor equipment that can directly measure the moisture in the air. The dewpoint sensor is usually collocated with the temperature sensor. One type of sensor in common use, including by United States ASOS sites, uses controlled refrigeration of a small mirror to produce condensation detectable by an infrared beam.

2.3.3. Wind. Wind is generally measured at a height of 10 meters (30 ft) above ground level. The direction that the wind is coming from, as well as its speed, are measured. The value is generally an average value during a period of the past 2 minutes, however gusts are considered over a period of the past 10 minutes.

2.3.4. Pressure. Pressure may be recorded or transmitted in one of many formats. They are listed below from most common to least common.

* Sea-level pressure (also known as SLP). This is the atmospheric pressure, corrected using r-values or r-factors based on observed temperatures during the past 12 hours (therefore cancelling out thermal contributions). This value is used on a widespread basis on surface maps worldwide. It is generally expressed in millibars or hectopascals. "Normal" sea-level pressure is 1013.2 mb.

* Altimeter setting (also known as ALSTG or QNH). The station pressure value (below) reduced to sea level. It is usually expressed in inches of mercury. A "normal" altimeter setting is 29.92 inches.

* Station pressure (also known as QFE). This is the actual pressure reading that would be indicated by a mercury barometer

---

Snowy tree crickets are said to be accurate thermometers. Count the chirps for fourteen seconds and add forty. The result is the air temperature in degrees Fahrenheit.

HUMOR BREAK
One of the classic questions in physics classes is, "How do you measure the height of a building with a barometer?" These are some of the answers that have appeared on answer sheets.
* Toss the barometer off the roof, then use the formula $S=0.5at^2$ to figure the height.
* Tie the barometer to the end of a rope, lower it to the street, then determine the length of the rope.
* Tell the superintendent that you will give him the barometer if he tells you the height.
* Sell the barometer and buy a tape measure.
* Use the barometer as a paperweight while examining the building plans.
* Pack the barometer with explosives, take it to the top of the building, and set it off. Listen for the news report that says, "There was an explosion today on top of the X-foot skyscraper..."
* Beat the barometer on the foundation until the building comes crashing down. The height will equal zero.

(i.e. if we were at a height of 18,000 ft, the station pressure would probably be 520 mb or 15.38 inches of mercury). It is the a direct, uncorrected measure of pressure. Although it is rarely transmitted, it is the basis for finding further pressure values like station pressure and altimeter setting. Station pressure may be expressed in either inches or millibars. It is also the value to use when dealing with physical equations.

2.3.5. Visibility. Visibility is a measure of opacity of the atmosphere *in a particular direction*. It indicates the distance that a normal object under daytime illumination can be seen and recognized. If the visibility is 3 miles, then any objects that are further than 3 miles away are unlikely to be reliably identified. The **prevailing visibility**, which is the value used on nearly all surface observations, is defined as the greatest visibility met or exceeded throughout at least half of the visible horizon.

2.3.6. Weather and obstructions to vision. Weather, such as rain or snow, and obstructions to vision, such as fog, haze, or smoke, are noted. Continuous or intermittent precipitation is distinguished from showery precipitation by the fact that showers are produced by convective clouds (cumulus or cumulonimbus) and tend to start or stop abruptly. Automated stations detect scintillation from an infrared beam in order to determine the precipitation or obscuration type.

2.3.7. Cloud height. Cloud height is the base of the cloud layer, usually measured to the nearest hundred or thousand of feet. If the layer and the layers below it add up to more than half of the sky, the higher layer's height is considered to be the ceiling. If the sky is solid overcast, for example, the ceiling is considered to be the height of the base of the overcast layer.

2.3.8. Cloud amount. Cloud amount is measured based on summation. If more than one layer is present, higher layers are the sum of the visible portion of the higher layer plus all visible lower layers. Total cloud amount is the amount of sky covered by clouds. Automated stations are capable, to a limited extent, of measuring cloud amount. They generally are unable to measure high cirrus clouds.

Abbv	Meaning	Description
CLR	Clear	No visible clouds
SCT	Scattered	1/8 to 4/8 coverage
BKN	Broken	5/8 to 7/8 coverage
OVC	Overcast	Total coverage

During the early 1880's the Canadian Pacific Railway and its telegraph lines were pushed across western Canada allowing the establishment of telegraphic weather reporting stations and many climatological and precipitation reporting stations. In eastern Canada during the summer of 1881 the [meteorological] service began to issue forecasts at midnight so that these might be published in the morning newspapers and also be displayed at telegraph stations as soon as they were opened each morning. Another innovation of the early 1880's was the dissemination of weather predictions by means of display discs attached to railway cars. The signal word to be displayed was telegraphed each day at about 1 a.m. to the railway agents who would change the signs on cars each morning in an attempt to provide a reliable weather prediction service for the farming community along the lines of the railway. However, through neglect, the local train hands did not always keep the signal discs up to date, and this arrangement had to be dropped after a decade or so.

MORLEY K. THOMAS, "A Brief History of Meteorological Services in Canada," 1971

Figure 2-3. Stratocumulus layer in north Texas.

Figure 2-4. Cumulonimbus cloud in southwestern Arizona.

Figure 2-5. Interestingly, a large number of basic textbooks incorrectly use weak cumulonimbus clouds to illustrate nimbostratus. Nimbostratus is a dull, featureless cloud, as in the photo seen above, and rain or snow usually falls from it. It is often obscured by low stratus ("scud") clouds.

2.3.9. Cloud type. Unfortunately, observation of cloud type is rapidly becoming neglected in the headlong rush to automate all observations. Up until 1991, cloud types were not only a common part of observations, but useful comments about moderate or towering cumulus or altocumulus castellanus in a certan direction appeared in hourly weather reports. Now that humans have largely been eliminated, this is no longer the case. At some ASOS stations, a human observer may code cloud types every three hours, but this sort of data is more likely to come from Ecuador and Ethiopia than from Vermont and Maine.

Nevertheless, having a knowledge of cloud types is still essential in interpreting satellite imagery, understanding conceptual weather processes, and in reading the sky. As I write this paragraph, being able to tell the difference between stratocumulus and cumulus on a satellite photo made the difference in creating a successful severe weather forecast just three days ago. I was able to tell where the atmosphere was strongly capped during the late afternoon hours. And knowing the appearance of a cloud on a satellite image starts with knowing how it looks visually, standing on the ground.

* *Stratus*. Also known as "scud" and "pannus" in poor weather conditions, stratus is an amorphous, very low cloud, with a fairly uniform base. It may precipitate drizzle, snow, or snow grains. Rain may be present, but it is always caused by other clouds in conjunction with the stratus. When the sun is visible through stratus, its outline is clearly discernible.

* *Stratocumulus*. Grey or whitish, or both grey and whitish, patch, sheet or layer of cloud which almost always has dark parts, composed of tessellations, rounded masses, or rolls, which are non-fibrous (except for virga), and which may or may not be merged. Stratocumulus may sometimes be confused with altocumulus. In very cold weather, stratocumulus may produce abundant ice crystal virga.

* *Cumulus*. Detached clouds, generally dense and with sharp outlines, developing vertically in the form of rising mounds, domes, or towers, of which

the bulging upper part often resembles a cauliflower. The sunlit parts of these clouds are mostly brilliant white; their base is relatively dark and nearly horizontal. Sometimes cumulus is ragged.

Cumulonimbus. A heavy and dense cloud, of considerable vertical extent, in the form of a mountain or huge towers. By convention, it is the only cloud that can produce thunder, lightning, and/or hail (but it doesn't necessarily do so). At least part of its upper portion is glaciated (smooth, fibrous, or striated), and it often spreads out in the shape of an anvil or vast plume. Under the base of this cloud, which is often very dark, there are frequently low ragged clouds either merged with it or not, and precipitation, sometimes in the form of virga in dry air. Aloft, there is the horizontal spreading of the highest part of the cloud, leading to the formation of an "anvil". Often, strong upper-level winds blow the anvil downwind in the shape of a half anvil or vast plume.

Nimbostratus. Dense, grey cloud layer, often dark, the appearance of which is diffused by more or less continuously falling rain or snow which, in most cases, reaches the ground. It is thick enough throughout to block the sun. Nimbostratus is generally an extensive cloud, the base of which is frequently partially or totally hidden by ragged scud clouds (stratus). Care must be taken not to confuse these with the base of the nimbostratus. Scud clouds and the nimbostratus may or may not merge. Also, nimbostratus can be distinguished from thick stratus by the type of precipitation it produces (see chart). If hail, thunder, or lightning are produced by the cloud, it is then classified as cumulonimbus.

Altostratus. Greyish or bluish cloud sheet or layer of striated, fibrous, or uniform appearance, totally or partially covering the sky, and having parts thin enough to reveal the sun at least vaguely, as if through ground glass. Altostratus prevents objects on the ground from casting shadows. If the presence of the sun or moon can be detected, this indicates altostratus rather than nimbostratus. If it is very thick and dark, differences in thickness may cause relatively light patches between darker parts, but the surface never shows real relief, and the striated or fibrous structure is always seen in the body of the cloud. At night, if there is any doubt as to whether it is altostratus or nimbostratus when no rain or snow is falling, then, by convention, it is called altostratus. Altostratus is never white, as thin stratus may be when viewed more or less towards the sun.

JOHN OPIE

Cloud nomenclature
For many centuries, it was thought that clouds were too ambiguous to be scientifically identified and classified. However in 1802, Luke Howard, a dedicated English amateur meteorologist, followed the lead of Swedish taxonomist Linnaeus and developed a scheme for categorizing different types of clouds. The scheme used four Latin names "cumulus" (heap), "stratus" (layer), "nimbus" (rain), and "cirrus" (curl), and allowed for mixtures of the words as needed. Howard presented this scheme in his "Essay on the Modification [Classification] of Clouds", presented to London's philosophically-oriented Askesian Society. In the decades ahead it was adopted by meteorologists worldwide.

* *Altocumulus*. White or grey, or both white and grey, patch, sheet, or layer of cloud, generally with shading, and composed of laminae, rounded masses, rolls, etc. which are sometimes partially fibrous or diffuse and which may or may not be merged. Most of the regularly arranged small elements usually have a visual width or between 1 and 5 degrees. Altocumulus sometimes produces descending trails of fibrous appearance (virga).

* *Cirrus*. Detached clouds in the form of white delicate filaments, or white or mostly white patches or narrow bands. These clouds have a fibrous appearance, or a silky sheen, or both. Cirrus is whiter than any other cloud in the same part of the sky. With the sun on the horizon, it remains white, while other clouds are tinted yellow or orange, but as the sun sinks below the horizon the cirrus takes on these colors, too, and the lower clouds become dark and/or grey. The reverse is true at dawn when the cirrus is the first to show coloration.

* *Cirrostratus*. Transparent whitish cloud or veil of fibrous or smooth appearance, totally or partially covering the sky, and generally producing halo phenomena. The cloud usually forms a veil of great horizontal extent, without structure and of a diffuse general appearance. It can be so thin that the presence of a halo may be the only indication of its existence.

* *Cirrocumulus*. Thin, white patch, sheet, or layer of cloud without shading, composed of very small elements in the form of grains, ripples, etc. merged or separate, and more or less regularly arranged. Most of the elements have a visual width of less than 1 degree. A rare cloud, cirrocumulus is rippled and is subdivided into very small cloudlets without any shading. It can include parts which are fibrous or silky in appearance but these do not collectively constitute its greater part.

Figure 2-6. Altocumulus cloud in north Texas.

Figure 2-7. Tangled mass of high cirrus clouds backdrops an American Airlines 727 landing at Dallas-Fort Worth International Airport.

2.3.10. Flying conditions. Because of the close association with meteorology and aviation, forecasters may sometimes run across distinctions for various flying conditions in discussions. These distinctions are classified according to ceiling and visibility. Many weather plotting software programs also color-code station plots according to flying conditions, which actually gives a handy way to instantly spot where humidities are high and lifted condensation levels are low. Definitions are as follows:

VFR — Visual flight rules. Ceiling above 3000 ft and visibility above 5 miles.
MVFR — Marginal visual flight rules. Ceiling between 1000 and 3000 ft or visibility between 3 and 5 miles. Conventional color code is blue.
IFR — Instrument flight rules. Ceiling less than 1000 ft or visibility less than 3 miles. Conventional color code is red.
LIFR — Low instrument flight rules. It's a subset of IFR defined as ceiling less than 500 ft or visibility less than 1 mile.
CAVOK — Not used in the United States. Visibility 10 km (7 miles) or more, no significant weather phenomena, no cloud below 5000 ft, and no cumulonimbus.

> Your hand, when held in a certain manner, spans 22 degrees, corresponding to the refraction angle of light through ice crystals. If cirrostratus is present, hold your hand fully outstretched at arm's length with the tip of your thumb covering the sun or moon. The tip of your little finger will touch the 22-degree halo if it is present. When held horizontally, the tip of your little finger will also touch a sundog if it is present.

REVIEW QUESTIONS

1. You have arrived in Madagascar and want to examine weather across the country. Would you look at METAR or synoptic reports?

2. You need to study a thunderstorm that moved through Tokyo, Japan and examine it hour-by-hour. Would you use METAR or synoptic reports?

3. While listening in to the control tower at the Oskhosh Airshow, you hear the controller giving the pressure in terms of inches. Is this sea-level pressure, altimeter setting, or station pressure?

4. Your mad scientist friend is fine-tuning his time machine and needs to know the air pressure for a set of physics equations. Should he use sea-level pressure, altimeter setting, or station pressure?

5. You are the weather observer at Chicago O'Hare. The visibility to the north is 1 mile, to the east is 2 miles, to the south is 4 miles, and to the west is 8 miles. What do you report for the prevailing visibility?

6. A station reports the total sky cover as being 3/8ths (scattered), made up of a stratocumulus layer with bases at 4,000 ft and tops at 6,000 ft. Is there a ceiling?

7. You are working at a weather station on an airfield. A pilot comes in and tells you that the cumulus clouds are topped at 8,000 ft and have bases at 3,000 ft. What is the layer height of the cumulus?

8. The top of a towering cumuliform cloud is showing soft tops but it doesn't have an anvil form. Is it a cumulus or cumulonimbus cloud?

9. A thick, gray, amorphous sheet of cloud covers the sky and completely obscures the sun. No rain is falling. What type of cloud is present?

10. A thin cloud covers the sky and people are marvelling at the large halo surrounding the sun. What type of cloud is present?

Hand analysis. *(Tim Vasquez)*

3 ANALYSIS

Receptivity allows one's thoughts to balance properly the daily stream of observational data, model data, and environmental clues, while enjoying the process of determining which clues are most critical. The more you love the forecasting and ingestion of data and environmental clues, the greater the success you will have. Get lost in the joy of the process!

ALAN R. MOLLER
NWS Forecaster, 2001

A forecast is nothing more than a guess unless the forecaster has a thorough understanding of what processes are happening in the atmosphere. Unfortunately this is the most neglected aspect of the meteorological science. A primary cause of this neglect is the widespread dependence on numerical models, often referred to as "meteorological cancer". A surprising number of forecasters use numerical models as a basis for their forecasts, rather than as a tool, and see current data as a curiosity that's best left to the computers.

A wise forecaster, however, puts the numerical models aside for a moment and looks at analysis tools: surface charts, upper air charts, satellite, radar, and more. Even the simple process of looking out the window provides a valuable scrap of analysis information. The forecaster's goal must be to sample the current state of the atmosphere in as many ways as possible in order to understand patterns and processes across the forecast area, even if they aren't immediately understood. The forecaster also tries to evaluate the trends of these patterns and processes. By using different tools and comparing features that are found against experience and common-sense meteorological knowledge, the forecaster has all the prerequisites for an accurate, scientifically-based forecast.

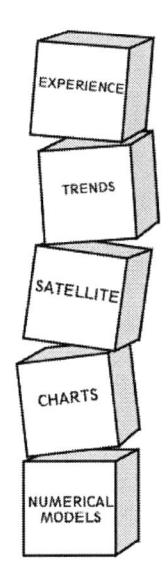

Figure 3-1a. Unbalanced forecast. Poorly skilled forecasters usually build from the bottom upward as shown here, forcing the next higher level to "fit" on top. Improper analysis of one product topples the entire forecast.

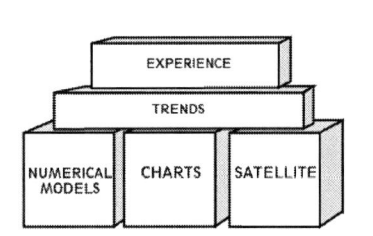

Figure 3-1b. A balanced forecast. Each forecast tool is used on its own merits then is synthesized with the other tools. Discrepancies and problems don't affect other tools.

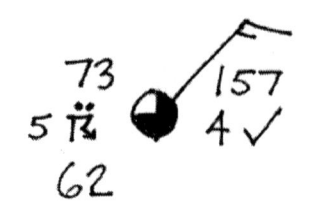

Figure 3-2. Surface plot model. This example shows a temperature of 73, a dewpoint of 62, with a visibility of 5 miles in a thunderstorm. The sea level pressure is 1015.7 mb, and the pressure tendency is 0.4 mb falling then rising. The sky cover (shading of circle) is 6/8ths, and the wind direction is northeast at 15 mph.

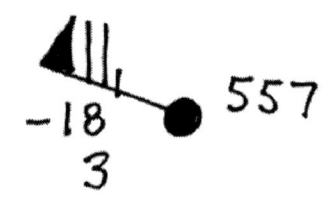

Figure 3-3. Upper-level plot model. This example is extracted from a 500 mb chart. The temperature is -18 deg C, and dewpoint depression is 3 deg C (therefore the dewpoint is -21 deg C). The height of the constant pressure surface is 5570 meters (the translation of these digits varies depending on the pressure surface being used; see text). The circle is shaded to indicate a dewpoint depression of less than 5 deg C (moist air). The winds are out of the west-northwest at 75 kts.

After a good 30 minutes to an hour looking at analysis products, an intermediate forecaster is often surprised to find that numerical model results either seem to fit in comfortably or stick out like a sore thumb.

3.1. CHART ANALYSIS

3.1.1. Station plots. Raw data is usually plotted on a map, station by station, as a series of numerical values. The most familiar example of this is when the Weather Channel shows a map with current temperatures across the United States. However when there is the need to express many different parameters for each station on a single map, such as wind, temperature, pressure, and so forth, international convention dictates that we use the *station plot model*.

The station plot model is an internationally-accepted style that condenses all the different types of numbers into visual groupings for each station, called *station plots*. This format allows a particular parameter to be easily seen across a map. For example, we can find the temperature simply by looking at the number on the top left side of each station plot. Pressure is found on the top right side of each station plot. By keeping each value in predictable "slots", it's surprisingly easy to scan a map for different types of data.

The most important part of the station plot is the *station circle*, which is a circle about 1/8th of an inch (2 to 4 mm) in diameter centered on the station location. The circle is shaded according to the total cloud cover present. A clear circle indicates clear skies, and a shaded circle indicates cloudy skies. An "X" in the circle indicates the sky is obscured by precipitation or other weather. Sometimes the station circle is not present and is represented by a cross, which highlights an automated station that can't assess the sky.

Radiating outward from the station circle is the *wind shaft*. It points into the wind, showing its direction of origin. At the tip of the wind shaft are feathers (barbs), representing the mean speed of the wind. A long barb indicates 10 kt, a short barb indicates 5 kt, and a flag-shaped barb indicates 50 kt. By counting the number of barbs, a forecaster can instantly interpret the wind speed to the nearest 5 kt. If there is no wind shaft or barbs and just a circle around the station circle, this indicates calm winds.

Temperature, visibility, and dewpoint are expressed in whole numbers. However there are pictograms for representing cloud type and precipitation, and even coded values for cloud height.

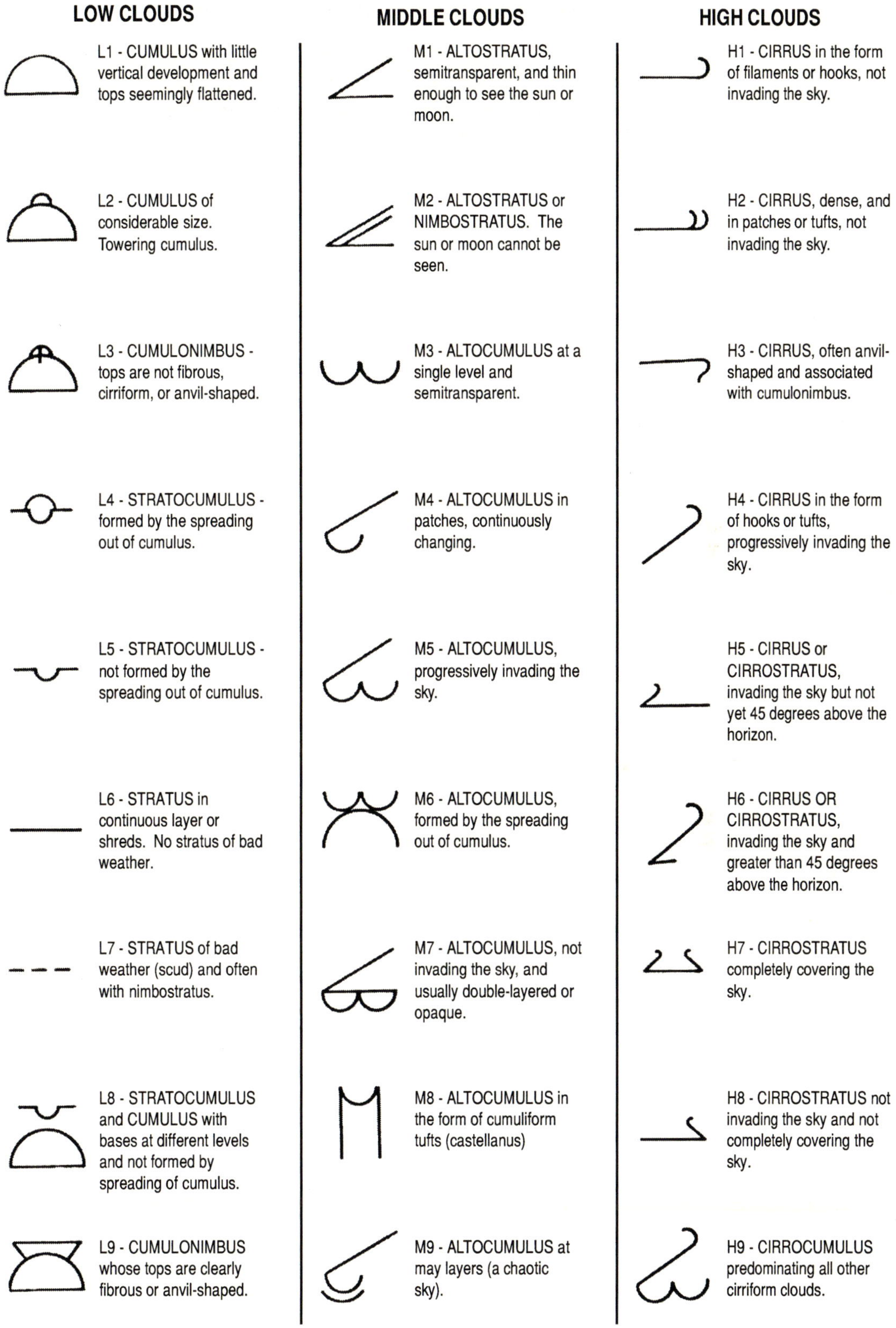

LOW CLOUDS	MIDDLE CLOUDS	HIGH CLOUDS
L1 - CUMULUS with little vertical development and tops seemingly flattened.	M1 - ALTOSTRATUS, semitransparent, and thin enough to see the sun or moon.	H1 - CIRRUS in the form of filaments or hooks, not invading the sky.
L2 - CUMULUS of considerable size. Towering cumulus.	M2 - ALTOSTRATUS or NIMBOSTRATUS. The sun or moon cannot be seen.	H2 - CIRRUS, dense, and in patches or tufts, not invading the sky.
L3 - CUMULONIMBUS - tops are not fibrous, cirriform, or anvil-shaped.	M3 - ALTOCUMULUS at a single level and semitransparent.	H3 - CIRRUS, often anvil-shaped and associated with cumulonimbus.
L4 - STRATOCUMULUS - formed by the spreading out of cumulus.	M4 - ALTOCUMULUS in patches, continuously changing.	H4 - CIRRUS in the form of hooks or tufts, progressively invading the sky.
L5 - STRATOCUMULUS - not formed by the spreading out of cumulus.	M5 - ALTOCUMULUS, progressively invading the sky.	H5 - CIRRUS or CIRROSTRATUS, invading the sky but not yet 45 degrees above the horizon.
L6 - STRATUS in continuous layer or shreds. No stratus of bad weather.	M6 - ALTOCUMULUS, formed by the spreading out of cumulus.	H6 - CIRRUS OR CIRROSTRATUS, invading the sky and greater than 45 degrees above the horizon.
L7 - STRATUS of bad weather (scud) and often with nimbostratus.	M7 - ALTOCUMULUS, not invading the sky, and usually double-layered or opaque.	H7 - CIRROSTRATUS completely covering the sky.
L8 - STRATOCUMULUS and CUMULUS with bases at different levels and not formed by spreading of cumulus.	M8 - ALTOCUMULUS in the form of cumuliform tufts (castellanus)	H8 - CIRROSTRATUS not invading the sky and not completely covering the sky.
L9 - CUMULONIMBUS whose tops are clearly fibrous or anvil-shaped.	M9 - ALTOCUMULUS at may layers (a chaotic sky).	H9 - CIRROCUMULUS predominating all other cirriform clouds.

Figure 3-4. Cloud type symbols.

Sea-level pressure is expressed in an unusual fashion: in tens, units, and tenths of a millibar. The hundreds place is omitted but is always assumed to be 9 or 10 (use 10 if the figure is below "500"). Therefore a figure of "975" equals 997.5 mb and "063" equals 1006.3 mb. This "500" rule does not work around extremely deep lows or very strong highs and the forecaster may have to choose the appropriate value subjectively ("577" could represent 957.7 mb in low pressure or 1057.7 mb in high pressure). Needless to say this ambiguity sometimes causes problems in contouring software.

Altimeter setting, in the rare cases that it is substituted for sea-level pressure, is also expressed in an unusual fashion: in units, tenths, and hundredths of an inch. The tens place is omitted but is always assumed to be 2 or 3 (use 3 if the figure is below "500"). Therefore a figure of "975" equals 29.75 inches and "063" equals 30.63 inches.

On upper-level charts, pressure representations are replaced by geopotential height values. These are always expressed as decameters (dam), therefore they represent height in thousands, hundreds, and tens of meters (e.g. "934" indicates 9340 m). The sole exception is the 850 mb chart which expresses height in whole meters, however 1000 must be added to this value (e.g. "524" is 1524 m). On charts at 300 mb and above, values may exceed 1000 dam, so plotted values of 500 dam and less must have 1000 dam added.

3.1.2. Chart analysis. Although glancing at a weather map containing nothing but data plots might be sufficient to get an idea of what's happening, it's sometimes necessary to draw isopleths (lines) to establish the shape of the most important fields. It's common for computer software or Internet-based graphics to provide these isopleths, however they do tend to smooth out important patterns or features, requiring re-analysis by hand.

The forecaster should sketch isopleths when time permits. This process does not simply accomplish an arcane task of drawing lines; rather it helps etch into the forecaster's memory all of the details of the atmosphere, and forces the forecaster to consider areas of unusual or unexpected readings. Surprisingly, as forecasters become experienced they rely increasingly on hand analysis because of its proven value in finding key features.

Isoplething is done by first selecting a type of data that is most relevant to the forecast (isopleths or isodrosotherms are a common choice), then selecting a contour interval that is established by convention. Ssome suggestions are shown in the surface and upper air chart descriptions that follow. The forecaster can also choose an interval that yields the best results.

Figure 3-5. Basic symbols for weather types. Variations are obtained by combining symbols. Multiple instances of precipitation symbols are drawn to indicate heavier occurrences (side by side, in triangle pattern, or in a diamond pattern).

For example some forecasters may want to analyze pressure fields every 2 mb instead of the conventional 4 mb.

It's important to resist the temptation to smooth the lines, fudging the isopleths where they don't accurately fit the data. The purpose of isoplething is to allow anomalies to stand out and be considered. Data values which are clearly erroneous may be circled by the forecaster and ignored, however all other data must be isoplethed accurately.

■ **Isobars.** The centers of highest and lowest pressure are of greatest interest as they indicate the center of important weather systems. Elongations, troughing, or ridging with time into an area are also of concern. Fronts always lie within pressure troughs, so the forecaster should expect to kink the isobars when

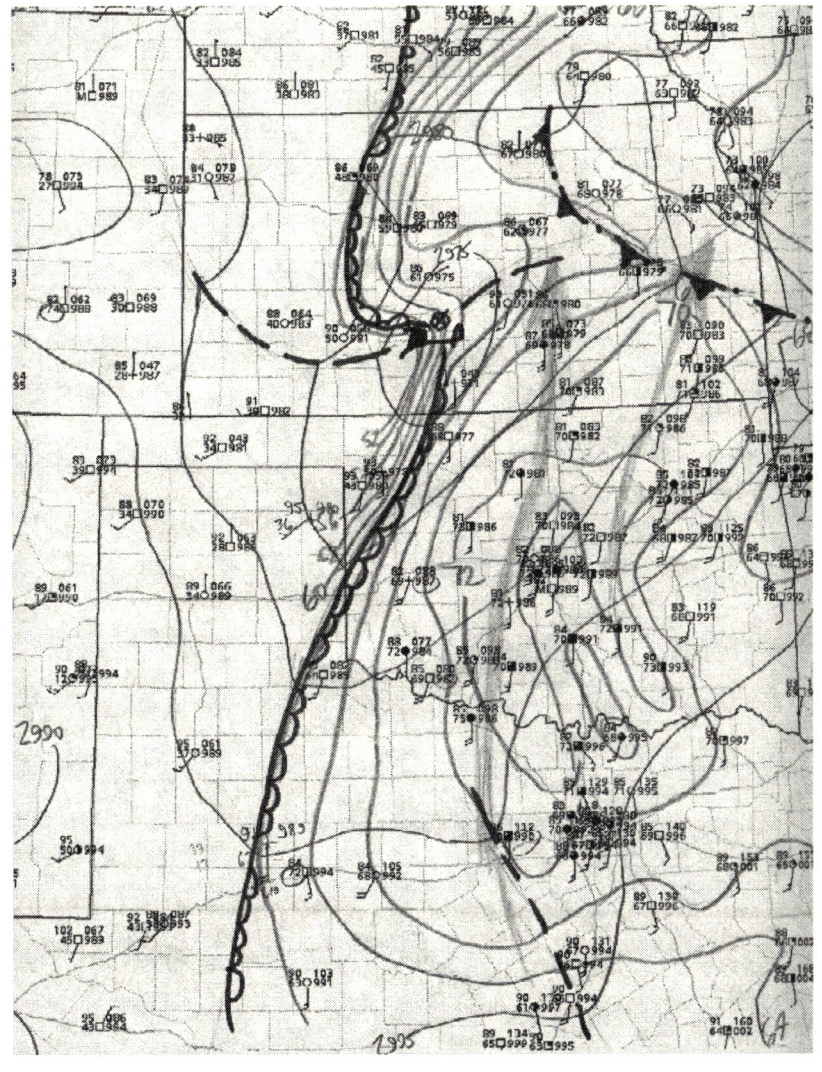

Figure 3-6. A computer plotted, hand-analyzed weather chart. The forecaster drew the isobars with a dark pencil, and outlined isodrosotherms with a green pencil. The result is a map showing the distribution of moisture around a dryline in Kansas, which allows an effort to determine where storms will develop.

Importance of map analysis
Many proponents of new technology believe that it is important to relieve forecasters of the "burden" of weather map analysis. From my perspective, this is terribly wrong! Performing map analysis is an essential component in diagnosis; it allows models of atmospheric behavior to be compared with the data. This is the way a forecaster forms an understanding of what is happening in the atmosphere. *Rather than freeing time to do science, taking map analysis away from forecasters minimizes their opportunity to function as practitioners of meteorological science.*

CHUCK DOSWELL
"The Human Element in Weather Forecasting," 1986

	Level (mb)		
Category	850	500	300/200
Weak	<20	<35	<55
Moderate	21-35	36-50	56-85
Strong	>35	>50	>85

Figure 3-3. A rough guide to the significance of wind speeds aloft.

drawing across a frontal boundary. Isobars are drawn in regular ink or pencil color from a base value of 1000 mb every 4 mb (sometimes 2 mb). Sometimes isobars can be constructed from altimeter setting values; these values are sensitive to terrain elevation and don't work reliably in mountainous areas, but they are slightly more common in the U.S. than sea level pressure, and since they aren't filtered by 12-hour temperature chages they are perfect candidates for measuring short-term pressure changes.

■ **Isotherms.** Generally it is not the actual patterns that are of interest, but the boundaries between warm and cold air where isotherms pack tightly. This can reveal important air mass differences. Wind flow perpendicular to an axis of isotherm packing suggests isentropic lift or sinking (which will be covered later in this book). Isotherms are drawn in red, usually every 2 or 5 degrees (either Celsius or Fahrenheit).

■ **Isodrosotherms.** By isoplething the dewpoint temperatures or mixing ratio (if available) the forecaster gets to visualize the surface moisture field. Before evaluating such fields its important to realize that moisture may have a depth that varies, which is an important consideration for thunderstorm development which depends on as much moisture possible throughout the greatest depth, and it may be important to look at soundings or charts at other levels above the ground when available. Nevertheless the axis of highest moisture may indicate where moisture is deepest. Isodrosotherms are drawn in green every 2 or 5 degrees (either Celsius or Fahrenheit).

■ **Isotachs.** By isoplething wind speed values, the shape of the strongest wind fields can be seen. This has the greatest value in the upper troposphere. By drawing isotachs on these charts, forecasters can keep abreast of the latest movements of jet stream winds, which have important effects on vertical motion (and weather) which will be covered in later chapters. Isotachs are drawn in purple at a value of every 20 kt from a base of 30 kt.

■ **Contours (isohypses or isoheights).** Reserved strictly for upper-level charts, contours connect values of equal geopotential height (the height of a pressure value). This forms a field functionally similar to an isobaric analysis, revealing low and high pressure areas. The regular pen or pencil color is used. Intervals and bases are as follows:

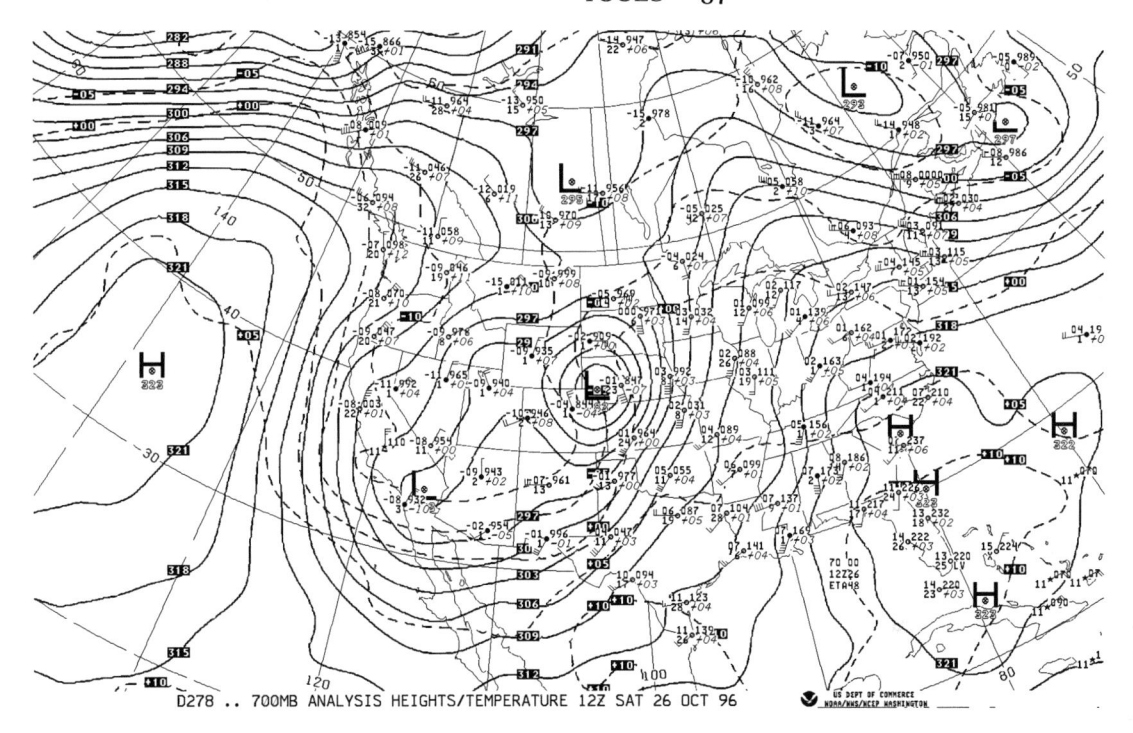

Figure 3-7. Typical 700 mb chart, showing conditions at roughly 10,000 ft MSL. The solid lines are height contours, and the dashed lines are isotherms. The "L" over colorado and southern California indicates low height areas; or simply "low pressure areas" at this altitude. Wind flows roughly parallel to the height contours with low pressure on their left and high pressure on their right. Tighter contours indicate stronger wind flow. It can be seen that winds are flowing strongly from north to south over northwest California and strongly from southwest to northeast over Oklahoma and Kansas.

Chart level	Base value	Interval
200/250/300 mb	900 dam	12 dam
500 mb	558 dam	6 dam
700 mb	300 dam	3 dam
850 mb	150 dam	3 dam

3.1.3. Surface chart. The surface chart gives us a representation of weather at the Earth's surface. Since it yields a much denser data field at more frequent intervals than upper-level charts, it is often used to keep close tabs on weather systems and air masses. Features such as pressure falls and frontal orientations can even suggest what is happening on upper-level charts.

3.1.4. 300/250/200 mb chart. Forecasters typically use <u>one</u> of these charts to track the jet stream, the steering flow, and other

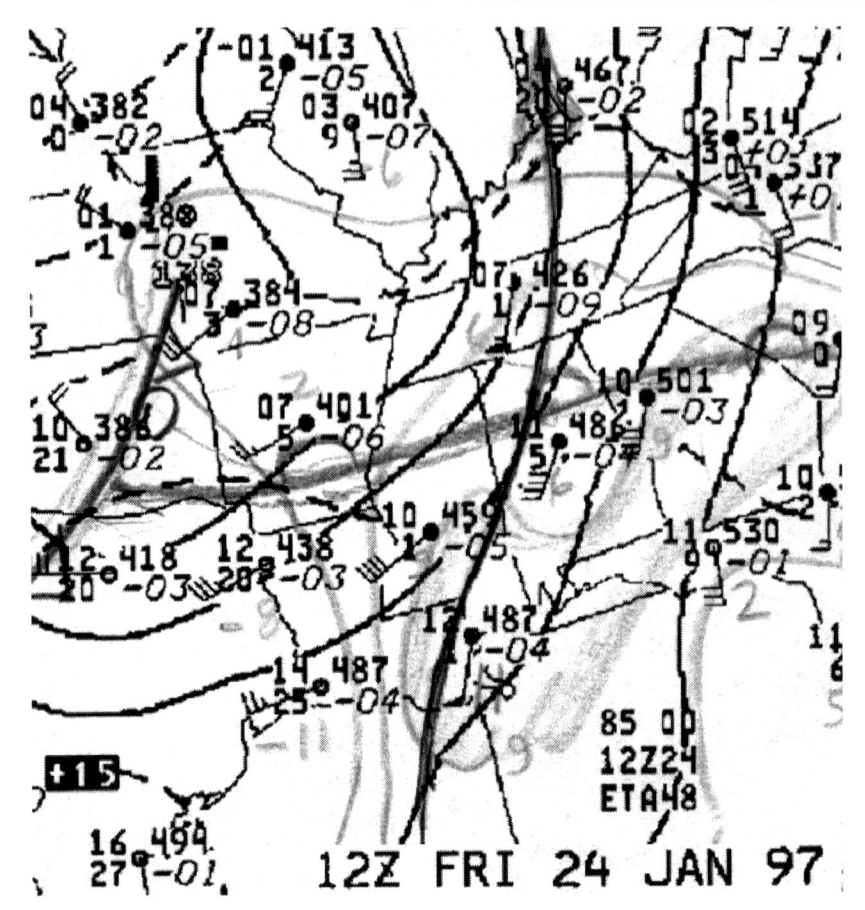

Figure 3-8. An 850 mb chart analyzed to forecast severe weather. Note the low-level jet drawn from New Orleans to Cincinnati and the warm front drawn from Atlanta to Texarkana. The analyst also converted the temperature and dewpoint depression to a dewpoint, writing the value at the lower right of each station plot, then sketched out the dewpoint field to determine the extent of rich moisture at 850 mb.

upper-level features. The chart that is chosen is the one that is in the highest levels of the troposphere but not above the tropopause. Since the tropopause height varies with the seasons, American forecasters use the 300 mb (30,000 ft) chart during the winter and the 200 mb (39,000 ft) chart during the summer. Some forecasters use the 250 mb (34,000 ft) during spring and fall.

Features typically analyzed are contours and isotachs, and the axis of jet streams are outlined with a red pen. The forecaster looks at the large waves and ridges, which outlines the long wave pattern (to be discussed in upcoming chapters) and gives an idea of the three-dimensional thermal structure across a continent.

3.1.5 500 mb chart. The 500 mb (18,000 ft) chart is a favorite of many forecasters, since it shows the broad-scale flow and jet streams seen on upper-level charts, yet it reveals some of the intricate details of low-level storm systems are still visible. The vast majority of model forecast charts show patterns at the 500 mb level.

Features analyzed include contours and isotachs. The axis of the jet stream is outlined in thick red pen. Short wave troughs (boundaries marked by strong cyclonic wind shifts) are outlined using a solid black line. Short wave ridges (features marked by strong anticyclonic wind shifts) are outlined using a black sawtooth line.

Absolute vorticity patterns (which will be discussed later) are sometimes drawn using automated methods every 2×10^{-5} rad/sec. These fields cannot be resolved by hand on a practical basis. Using vorticity patterns, forecasters will refine the positions of any drawn short waves, then annotate spots of highest cyclonic vorticity with an "X" (maximum) and spots of highest anticyclonic vorticity with an "N" (minimum). Areas of strong positive vorticity advection (PVA) are shaded in red, and areas of strong negative vorticity advection (NVA) are shaded in blue.

3.1.6. 700 mb chart. The 700 mb chart shows conditions at a height of about 10,000 ft. Many short wave disturbances are found at 700 mb. The 700 mb level winds are also used to provide a quick estimate of the movement of thunderstorm activity. During severe weather events, the 700 mb temperature patterns indicate the strength of the mid-level cap, which often prevents development or suppresses activity until later in the day.

Contours are drawn on the chart. Short wave troughs (boundaries marked by strong cyclonic wind shifts) are outlined using a solid black line. Short wave ridges boundaries marked by strong anticyclonic wind shifts) are outlined using a black sawtooth line.

3.1.7. 850 mb chart. The 850 mb chart helps complete the three-dimensional picture of low-level air masses. Boundaries and fronts can be located. Low-level warm air advection is associated with upward vertical motion, so areas where isotherms cross the height contours often indicate potential areas of cloud and precipitation development. Fronts aloft are also located.

The basic isopleth drawn on the 850 mb chart is contours. Fronts are located, placed on the warm side of strong temperature gradients.

Moisture is also outlined on the chart. For typical forecasting that does not involve severe weather, areas of

---

Computerized processing of weather data greatly reduces the manual task of drawing isopleths. However a meteorologist cannot passively depend on computer output. Instead, he or she must master both drawing lines and improving automatically drawn weather charts. A meteorologist is expected to prepare charts better than the computer does, even if not as fast. For example, computer output is known for the absence of discontinuities (fronts), poor resolution of mesoscale processes, and other shortcomings. Therefore, a professional meteorologist must know the classical methods of analysis, including those developed by the Norwegian and Chicago schools and numerous other investigators who advanced our knowledge of the detailed structure of the atmosphere. Thus, studying graphical manual drawing of weather charts is not obsolete in the era of automation.

DUSAN DJURIC,
"Weather Analysis", 1994

dewpoint depressions 5 C deg or less are outlined in thick green, signifying areas of possible clouds. For severe weather forecasting, contours for dewpoint temperature are drawn every 5 C deg. Since upper-air station plots, by convention, indicate dewpoint depression, the analyst may need to use quick arithmetic on each station plot to determine its dewpoint temperature.

3.1.8. 1000-500 mb thickness chart. The 1000-500 mb chart is simply a plot of the difference in height between the 1000 mb surface and the 500 mb surface. This difference is proportional to temperature. Therefore it is useful in helping to locate fronts in the lower and middle levels. It can also be used to locate upper-level jets, which lie along gradients of strong thickness contrast. Owing to its limited usefulness the 1000-500 mb chart is almost never drawn from scratch, but if a computer-produced chart is available it can be reviewed and used.

Lines of equal thickness are called isopachs, though few forecasters know this obscure bit of terminology. The contour interval is every 6 dam (60 m) from a base of 558 dam.

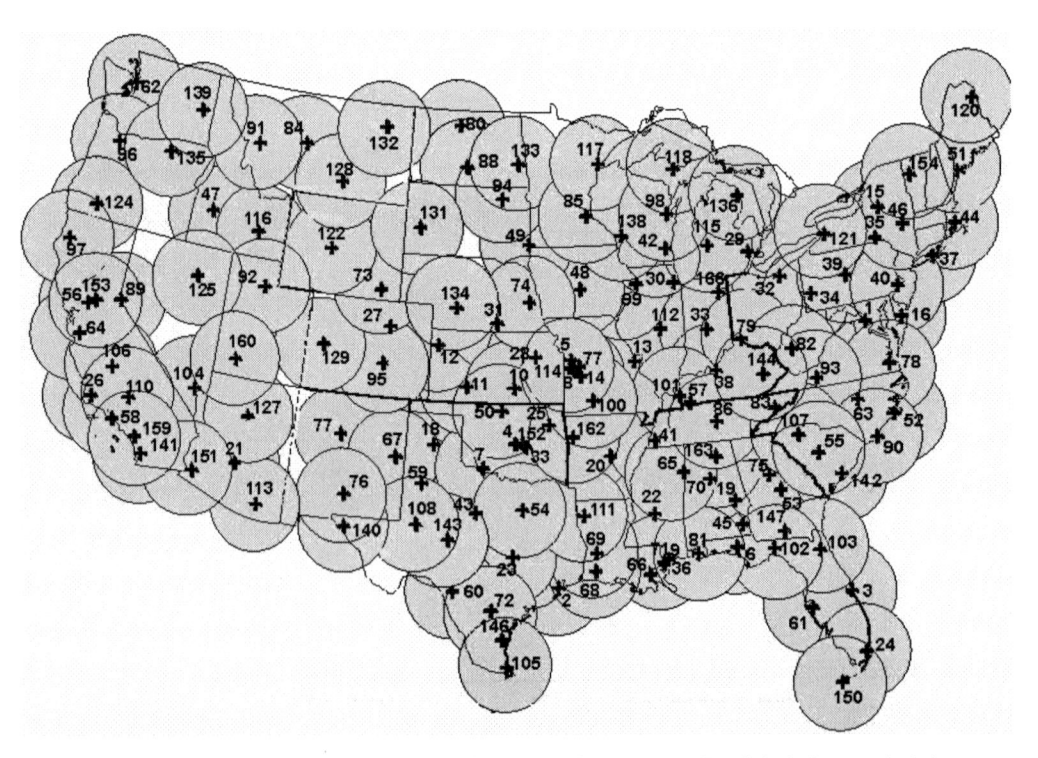

Figure 3-9. WSR-88D coverage in the conterminous United States as of October 2000. The digits indicate each site's sequence number. *(Courtesy NOAA/OSF)*

The estimated position of the polar front jet based on thickness gradient is drawn as a thick black arrow. The estimated position of surface fronts based on thickness gradients are drawn using standard conventions.

If surface isobars are available on the chart, areas of significant cold air advection are shaded with a blue pencil. Areas of significant warm air advection are shaded with a red pencil.

3.2. RADAR

From the 1950s to the 1980s, forecasters relied on the conventional radar network, which was composed of WSR-57 and WSR-74 units. Images were transmitted over dedicated or dialup telephone lines to RRWDS (Radar Remote Weather Display) systems or to proprietary display computers. The 1990s brought an end to this primitive state of technology. The United States radar network is now composed of over 100 WSR-88D (NEXRAD, or Next Generation Weather Radar) Doppler radar units. Images are distributed over weather datastreams and are widely available over the Internet.

3.2.1. Reflectivity. It is important to remember that weather radar is not designed to detect cloud droplets; it detects rain drops and hail. Therefore, small clouds will be invisible to the radar. Only the thickest of clouds will produce any detectable reflectivity values. Rainy precipitation areas and especially hail produce high values of reflectivity.

* *Thunderstorms.* Radar does not detect the cloud; rather it detects the rainy downdraft region of the storm (not the updraft nor the cumulonimbus towers specifically). Therefore any dangerous areas associated with the updraft may be on the edge of the storm or in an echo-free area altogether.

* *Severe thunderstorms.* Systematic study of radar images during the 1960s and 1970s revealed several indicators that are typically associated with severe storms. First, obviously, are higher reflectivity values, which are indicative of higher reflectivity from heavy rain and hail. Second is a core that has shifted to one side (frequently the south side), which occurs due to developing circulations within a storm. Third are reflectivity patterns that directly suggest strong wind circulations, such as hooks. Analysis of higher scans may show other indicators, such as a core aloft shifting to one side of the storm (an appearance caused by a very

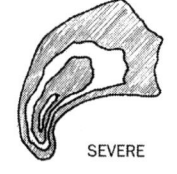

Figure 3-10. Reflectivity gradients. A non-severe storm tends to have low reflectivities, with the core well within the center of the storm. A severe storm often has high reflectivities, with a core shifted toward one side of the storm producing a strong reflectivity gradient on that side. It also shows reflectivity patterns that suggest a strong wind circulation (a hook, etc).

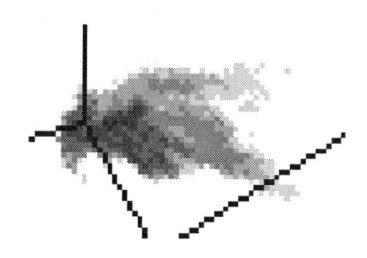

Figure 3-11. A tornadic storm in Texas, as seen on a reflectivity image. Note how the storm shows strong suggestions of a circulation, with a hook echo evident and a strong core near the south side of the storm.

42 • WEATHER FORECASTING HANDBOOK

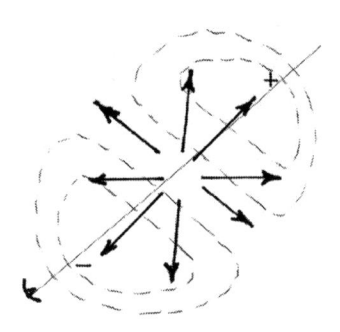

Figure 3-12a. Pure divergence shows a velocity couplet along the radar beam, with negative (inbound) flow on the near side of the circulation and positive (outbound) flow on the far side.

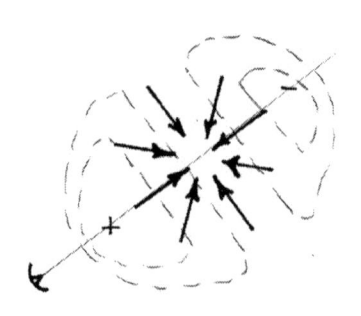

Figure 3-12b. Pure convergence shows a velocity couplet along the radar beam, with negative (inbound) flow on the far side of the circulation and positive (outbound) flow on the near side.

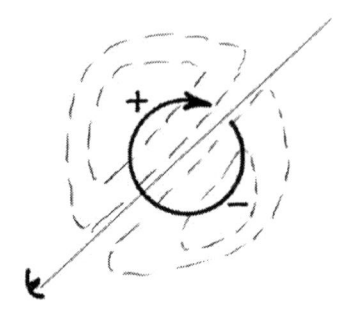

Figure 3-12c. Pure anticyclonic rotation shows a velocity couplet across the radar beam, with negative (inbound) flow on the right side and positive (outbound) flow on the left side.

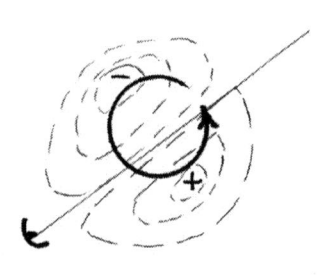

Figure 3-12d. Pure cyclonic rotation shows a velocity couplet across the radar beam, with negative (inbound) flow on the left side of the circulation and positive (outbound) flow on the right side.

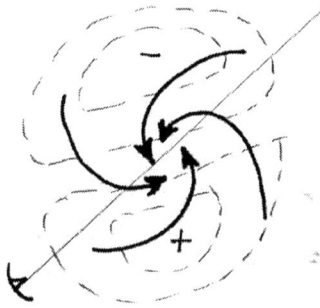

Figure 3-12e. Cyclonic convergence, as is often seen in severe storm mesocyclones, shows a couplet that is diagonal to the radar beam. Negative (inbound) flow is on the far left side of the beam and positive (outbound) flow is on the near right side.

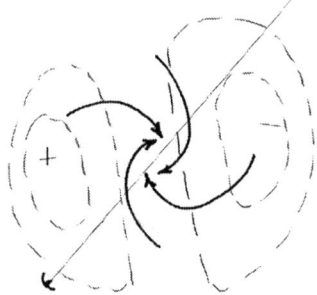

Figure 3-12f. Anticyclonic convergence, which we might see in a rare anticyclonic tornado, shows a couplet that is diagonal to the radar beam. Negative (inbound) flow is on the far right side of the beam and positive (outbound) flow is on the near left side.

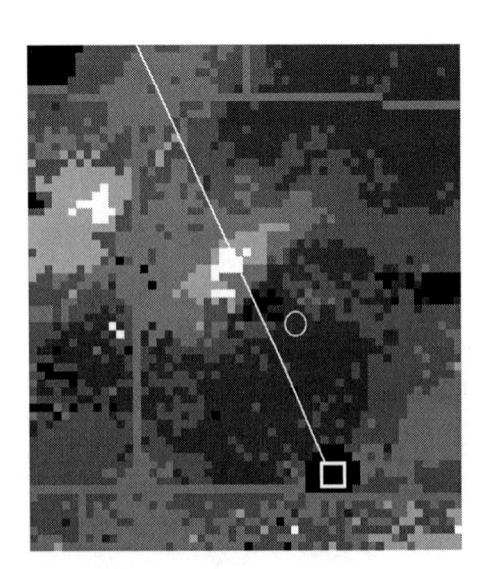

Figure 3-13. Tornado on velocity image. In this example we see a velocity image minutes before the March 28, 2000 Fort Worth tornado. The coloring has been set up so that high inbound (negative) velocities are bright, and high outbound (positive) velocities are shaded dark. The radar is at the square, and downtown Fort Worth is at the circle. Note that along the radar beam we see a couplet more or less aligned *along* the beam, with inbounds on the far side and outbounds on the near side. This indicates convergence. This convergence signature shortly evolved into a cyclonically-rotating convergence signature as a tornado touched down in Fort Worth. Four people were killed.

strong updraft which doesn't allow time for water droplets to form, producing what is known as a weak echo region, or WER, or in stronger cases a bounded weak echo region, or BWER).

3.2.2. Velocity. Decades of research has shown that the velocity signatures of storms and other atmospheric phenomena is of tremendous importance to forecasters. The late 1990's brought the widespread distribution of velocity images from radar data providers on the Internet, so learning how to interpret these images is a very useful skill.

* *Underlying principles.* It is important to remember that a radar site can only measure velocity in one dimension: *along* the radar beam. The radar has no way of measuring velocities that are perpendicular to the beam; it simply sees such movement as having zero velocity along the radar beam. Therefore it is up to forecasters to interpret velocity signatures.

* *VAD wind profiles.* At all times, NEXRAD radars collect data on winds throughout the entire scan volume. Even in clear air, dust, cirrus clouds, birds, and insects will scatter energy back to the radar, providing velocity information. The radar processing system measures the average winds at each height and produces a product showing this wind profile. Although this data can be of questionable value, the availability every few minutes around the country can be of significant value in keeping tabs on winds aloft when rawinsonde data isn't available.

* *Analyzing thunderstorms.* Forecasters look closely at thunderstorms using the velocity channel (typically with the lowest, or "base level" scan). Strong velocity signatures are indicative of strong winds within the storm. Couplets, which are a pair of opposing signs of wind, may indicate a strong circulation or a motion within the storm. By noting the alignment of the couplet with the radar beam, it can quickly be established whether the couplet indicates rotation and/or divergence/convergence (see illustrations).

3.3. WIND PROFILERS

Wind profilers are low-power radar devices that transmit a signal vertically into the atmosphere and analyze the backscattered signal to determine wind speed and direction throughout the column. The principle behind the system is that the turbulent mixing of air through the column, which yields varying indices of refraction, reveals important properties of the wind. The system operates at 6 kW at 404 MHz, using a flat

44 • WEATHER FORECASTING HANDBOOK

Figure 3-14. Profiler station equipped with RASS equipment.
(Courtesy NOAA/FSL)

Figure 3-15. NOAA Profiler network as of 2002.
(Courtesy NOAA/FSL)

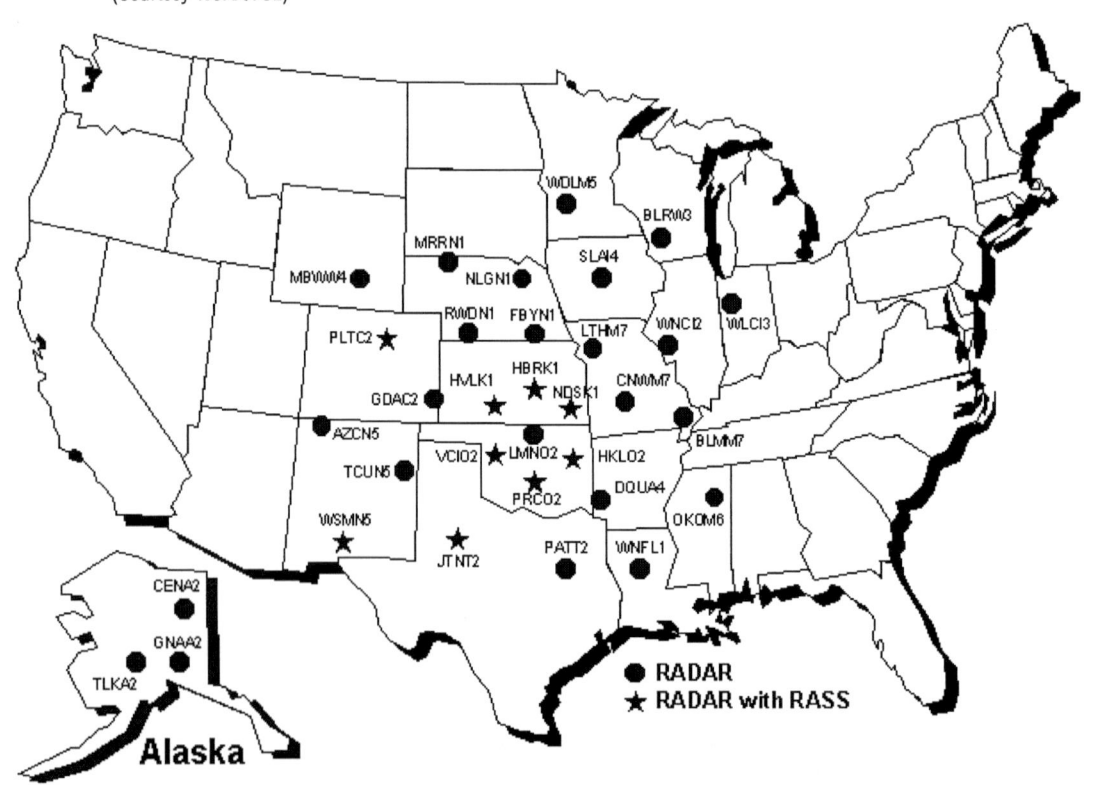

mesh antenna about the size of a tennis court. The main advantage of this antenna system is that it has no moving parts and requires less maintenance.

Wind profilers have proven to be essential elements in the upper-air network. Their value was proven on May 3, 1999 when a small-scale jet streak not resolved by the rawinsonde network or numerical models emerged over New Mexico and became associated with the development of Oklahoma's worst tornado outbreak.

Often collocated with wind profiler stations are RASS (radio acoustic sounding system) facilities. The signal is generated by a loudspeaker that projects an acoustic sound of about 850-900 Hz upward into the column of air above the station. Needless to say, these devices can be heard around the station, so foam insulation is used to reduce noise pollution. The speed of sound varies with temperature, and this can be sampled to construct a profile of temperature with height. The main drawback is that the vertical resolution of RASS data is nowhere near as good as balloon-launched soundings, and data may be unavailable above 10-15 thousand feet without higher clouds, but the trends over time in each layer can provide valuable information.

The value of RASS has not yet been determined, though it does show some promise in sampling the strength of capping inversions that inhibit thunderstorm development.

3.4. SATELLITE

Most American weather agencies use the GOES satellites, which are geostationary (rotating exactly with the Earth) 22,300 miles above the Equator.

3.4.1. Coverage. This section will review important weather satellites.

* GOES *(America, various positions)*. This series of satellites evolved from the prototype ATS series launched in 1967. GOES-10 (K), launched in April 1997, is used for the GOES-WEST (105 deg W) position. GOES-8 (I), launched in April 1994, is used for the GOES-EAST (75 deg W) position. GOES-11 (L) is on standby to be launched in case of a failure. What about GOES-9 (J)? It was was launched on May 23, 1995 but was replaced by GOES-10 in 1997 due to a mechanical problem. It is still online as an emergency spare.

* METEOSAT *(Europe, 0 deg)*. The modern series of METEOSAT satellites were launched in 1988. Most of the data

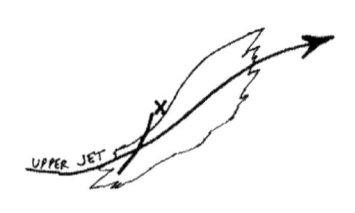

Figure 3-16. The baroclinic leaf pattern is occasionally seen on satellite imagery. This is associated with the first stages of developing frontal systems. The upper-level jet position is usually as shown here, with a short wave extending along the back side of the system.

Figure 3-17. Comma cloud. Baroclinic leafs often develop into large frontal systems manifesting theirselves on satellite imagery as "comma clouds". This conceptual model of a comma cloud shows the approximate location of surface fronts and the upper level jet.

(except for reprocessed images) are considered commercial property and are not shared. METEOSAT transmits 1 km vis and 3 km IR.

* *GMS (Far East, 140 deg E).* This Japanese series of satellites was launched in the 1970s. Images are freely distributed. The latest satellite, GMS-5, uses 1.25 km vis and 5 km IR/WV.

* *INSAT (Asia, 83 deg E).* Launched in 1990, India soon began encrypting the data, but an agreement in late 1997 allowed free data exchange again. Plans are underway to launch more INSAT satellites during the early 2000s. The satellite transmits 2 km vis and 8 km IR.

* *FENG-YUN 2 (Asia, 108 deg E).* The first satellite exploded on the launchpad during fueling on April 2, 1994, killing 1 and injuring 20. The second satellite began service late in 1997, but imaging problems began in 1998. Another launch is expected in May 2000. The FY2 series transmits 1.25 km vis and 5 km IR/WV.

* *GOMS (Asia, 77 deg E).* Known as Elektro, this Russian satellite was proposed in the 1970's but was finally launched in 1994. It transmits 1.25 km vis (the visible imagery package is nonfunctional now) and 6.25 km IR.

* *Polar orbiters.* Occasionally, polar orbiters (such as the DMSP and NOAA series) are used. These satellites are in sun-synchronous orbit (actually orbiting the earth over the poles at a significant speed) and are of great help where geostationary satellites are inoperative or can't see (such as in polar regions).

3.4.2. Visible. The visible channel measures light using the same wavelengths as the human eye. Naturally, objects with high albedo (thunderstorms and snow) are brighter than objects with low albedo (forests, water surfaces, etc). Visible imagery is only available during daylight hours.

3.4.3. Infrared. Infrared channels measure the radiation at 12.8 microns. Brightness is heavily a function of temperature. Therefore cold objects (clouds, polar regions, etc) are brighter than warm objects (ground, water, etc). Stratus, fog, and other clouds near the ground often blend in with the ground since they have the same temperature, but one technique to detect such clouds is to note the obscuration of lakes (the lakes will "disappear" when stratus moves over them).

TOOLS • 47

3.4.4. Water vapor. This channel measures the amount of radiation emitted by an object at 6.7 microns. Water vapor heavily absorbs radiation at 6.7 microns, and absorption is greatest in the mid/upper levels of the troposphere, between 600 and 250 mb (15,000 to 34,000 ft). Therefore, where patterns are white we can assume that radiation from the lower levels of the atmosphere are being absorbed by mid/upper level moisture, and thus, mid/upper level moisture is present. Where areas are dark we assume that low-level radiation has not been absorbed by mid/upper level moisture, and thus, the mid/upper levels are dry. Note that because absorption is greatest in the mid/upper levels, it is impossible to draw conclusions about low-level moisture. Water vapor imagery brightness is also a function of the low-level temperature (the source of the low-level radiation); in cold regions, little radiation is emitted and we cannot make conclusive interpretations about water vapor imagery.

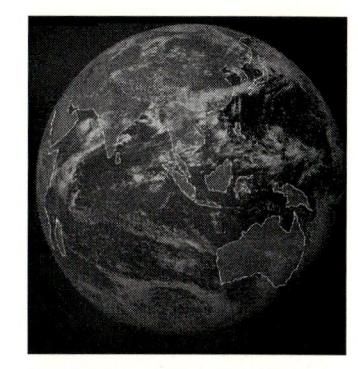

Figure 3-18. Chinese Feng-Yun geostationary weather satellite on June 21, 1997 delivers its first image. It continues to operate, but images have been sporadic since 1998.

3.5. SOUNDINGS

Although we are accustomed to analyzing weather horizontally using weather maps, we also need to look at the atmosphere vertically above a given station. The chart that we use to get this vertical picture is the "sounding". A sounding can be as simple as a graph of temperature versus height, though forecasters usually depend on more formal versions of such a diagram.

This section will examine four historical diagrams (emagram, Stüve, pastagram, and aerogram) and one international diagram (tephigram) that a forecaster may on occasion run into. Readers exclusively doing North American forecasting may prefer to skip those sections and proceed to the Skew-T section.

3.5.1. Source of data. Soundings are taken twice a day at 92 radiosonde stations across the United States and 32 in Canada. The sensor, about the size of a milk carton, is launched on a weather balloon. During its 45-minute ascent, the sensor sends a telemetry signal back to the station containing pressure, temperature, and humidity data. This data is collected by radiosonde processing software and is formatted into an internationally-standardized radiosonde report format known as TEMP (WMO FM-35 or TTAA) format. It is then distributed through weather networks worldwide. Although this numerical data can be decoded by hand and directly plotted on charts by skilled forecasters, many software programs and Internet sites relieve the forecaster of this burden. The values from the TEMP

Figure 3-19. Russian GOMS/Elektro geostationary satellite image on June 21, 1997 at 12:02 UTC (comparable date/time to the Feng-Yun image above), showing western Asia. Excellent coverage of the Indian Ocean is provided, but the visible imagery package no longer works.

format are also used to produce upper air charts (covered earlier).

3.5.2. Emagram. The emagram was created in 1884 by electromagnetism expert Heinrich Hertz to eliminate tedious calculations of adiabatic changes. It represents the very first chart useful for plotting atmospheric soundings. The coordinate system is linear temperature (X-axis) with quasi-linear pressure (Y-axis) and it provides straight lines for both.

3.5.3. Pseudoadiabatic (Stüve) diagram. Developed in 1927 by G. Stüve, the Stüve diagram is an improvement on the emagram. It is visually the simplest and most intuitive of all thermodynamic charts. It is almost identical to the emagram, however it has the added bonus of straight potential temperature lines, making three sets of perfectly straight lines. The Stüve diagram was widely used in the United States until the 1960s, when it was replaced by the skew-T.

3.5.4. Pastagram. Created in 1945 by J.C. Bellamy, this unusual diagram has a coordinate system of linear entropy (X-axis) and logarithmic pressure (Y-axis). It is similar to the Stüve diagram but features skewed temperature lines.

3.5.5. Aerogram. Created in 1935 by A. Refsdal. This was the forerunner to the skew-T diagram. Its coordinate system is linear temperature (X-axis) and logarithmic pressure (Y-axis). Equal areas represent equal amounts of work, and it was useful for determining integrated values of instability.

3.5.6. Tephigram. Created by Sir William Napier Shaw in 1922, the tephigram uses a coordinate system of linear reverse temperature (X-axis) and linear potential temperature (Y-axis). In the late 1940s the diagram was modified by rotating it about 45 degrees counterclockwise so that pressure lines would be horizontal, and the temperature coordinate was no longer reversed. The tephigram is structurally very similar to the skew-T but has pressure lines that curve slightly instead of being straight horizontal. This attention to detail makes it perfect for exacting physical calculations. The tephigram is still a favorite among European users, particularly in the United Kingdom.

3.5.7. Skew-T. The skew-T diagram, a variation of the emagram, was developed in 1947 by N. Herlofson and was quickly embraced by the U.S Air Force, gradually gaining acceptance in the National Weather Service. It's by far the most common chart used in the United States and Canada, and is what will be

Figure 3-20. A radiosonde package is a little bigger than a soda can (right). These little devices are launched twice a day at hundreds of weather stations worldwide, forming the backbone of modern weather forecasting.

used throughout this book. The diagram is very similar to the modern tephigram, except that pressure lines are perfectly horizontal to make it easier to estimate altitudes. It is composed of the following items.

* *Pressure lines.* Pressure (height) lines run left-right on the chart. To plot the position of a parcel of air, we put a dot higher on the chart if it is at a higher height. The lines are marked in millibars, but many diagrams also have a height scale to the far right which indicates feet and meters.

* *Temperature lines.* Temperature lines are straight solid lines running from lower left to upper right on the chart. The lines are marked in degrees Celsius. They are used to determine the temperature or dewpoint of a parcel at a given height.

* *Mixing ratio lines.* Mixing ratio lines are straight dashed lines running from lower left to upper right on the chart. They are marked in g/kg. These lines are used to determine the mixing ratio or saturation mixing ratio of a parcel at a given height.

* *Dry adiabat lines.* These are slightly curved lines running from lower right to upper left. They are marked in units of potential temperature in degrees Kelvin. A parcel always follows this line if it is dry (not saturated) and is rising or sinking.

* *Wet adiabat lines.* These are curved lines running from lower right to upper left. They are marked in units of potential temperature in degrees Celsius. A parcel always follows this line if it is saturated (condensing) and is rising. Sinking parcels only follow the dry adiabat lines.

3.6. HODOGRAPH

The hodograph is a trace of wind direction and speed with height at a given station plotted on an azran diagram. The information usually comes from radiosonde reports, but radar wind profiles and profiler data can be used to construct a hodograph trace. Hodographs are instrumental in measuring vertical shear within the atmosphere and determining what type of thunderstorms will develop.

The wind speed and direction at each level in the atmosphere above a certain station is plotted. Each data point is labelled with its height in km or feet. All the data points are

Radiosondes
Much interest has been shown recently in developing meteorographs which are capable of transmitting by radio a record of conditions which they encounter as they ascend into the atmosphere attached to small free balloons. A number of such types have been developed within the last few years, some of which have yielded very satisfactory results. This method of obtaining upper air information promises to supplant the use of airplanes in the very near future.

GEORGE F. TAYLOR
"Aeronautical Meteorology",
1938

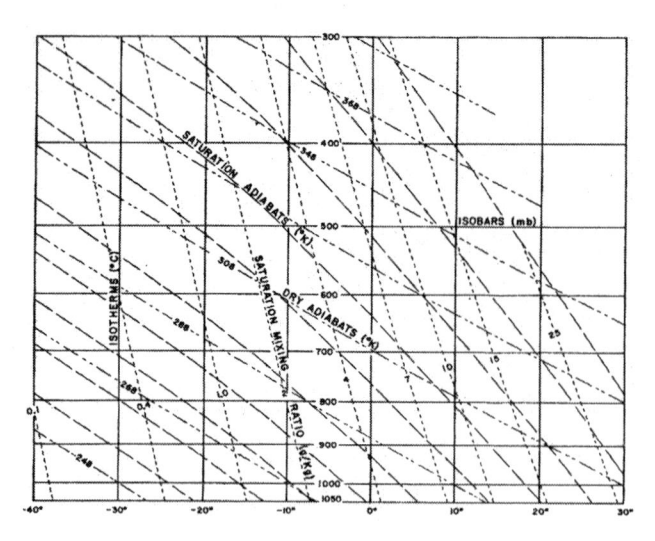

Figure 3-21a. Emagram. Developed by Heinrich Hertz in 1884, the emagram is the basis for all thermodynamic diagrams that were developed afterward. (From *The Use of the Skew-T Log-P Diagram in Analysis and Forecasting*, 1979, AWS)

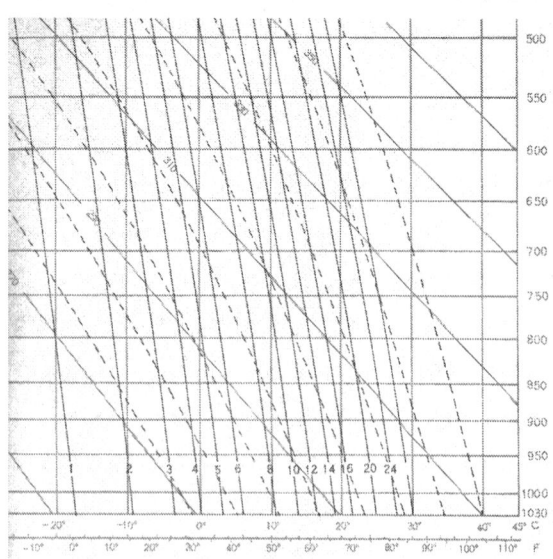

Figure 3-21b. Stüve (pseudoadiabatic) diagram. This diagram was in heavy use by the Weather Bureau during much of the 1940s and 1950s. Note how the temperature and pressure coordinates form a gridlike pattern. The potential temperature line is perfectly straight (converging towards the upper left), making it easy to differentiate it from an emagram.

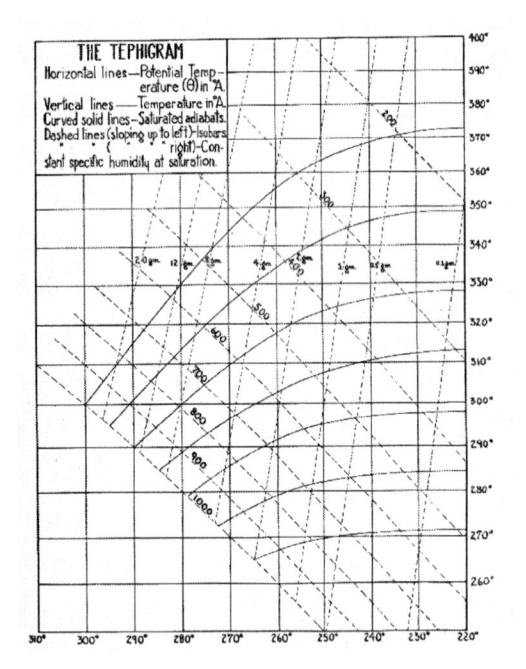

Figure 3-21c. Early tephigram. This is the classic form of the tephigram as it existed up until the late 1940s. Note how the coordinate system is linear **reversed** temperature (X-axis) with linear potential temperature (Y-axis), both labelled here in Kelvin. (From *Aeronautical Meteorology*, 1938, Taylor)

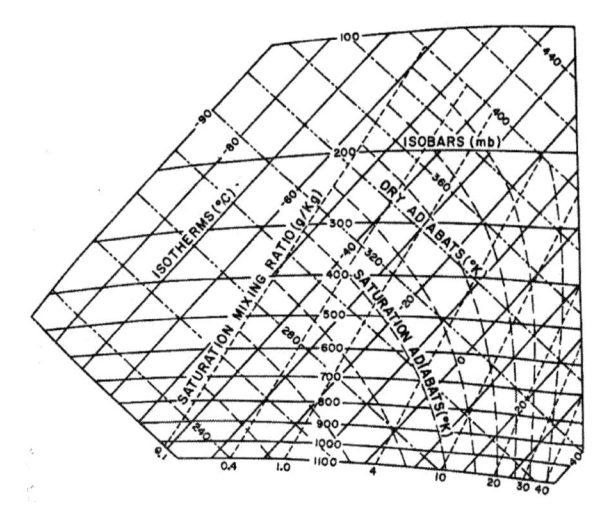

Figure 3-21d. Modern tephigram. The modern tephigram grew popular during the 1950s. This is simply the early tephigram rotated counterclockwise 45 degrees to make the pressure lines vertical. By looking at the early tephigram and turning the page in this fashion, the reader can see the change. (From *The Use of the Skew-T Log-P Diagram in Analysis and Forecasting*, 1979, AWS)

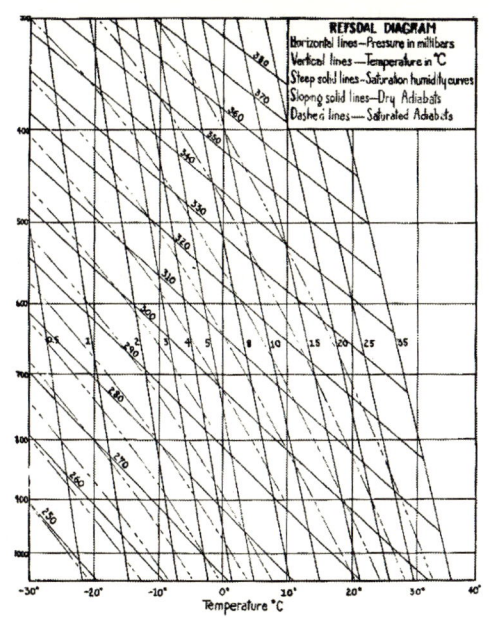

Figure 3-21e. Refsdal (aerogram) diagram. This is the forerunner to the skew-T diagram. Its coordinate system is linear temperature (X-axis) and logarithmic pressure (Y-axis). (From *Aeronautical Meteorology*, 1938, Taylor)

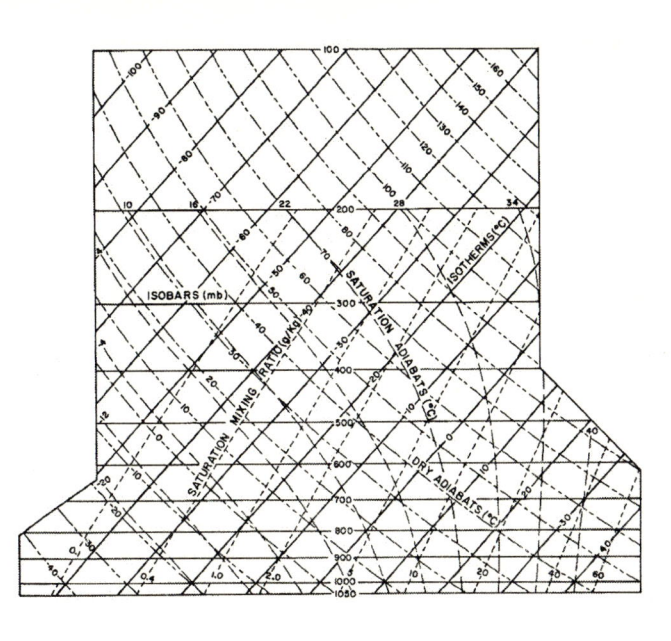

Figure 3-21f. Skew-T log P diagram. This is the most common diagram used in the United States and Canada. Complete descriptions are on the following pages.

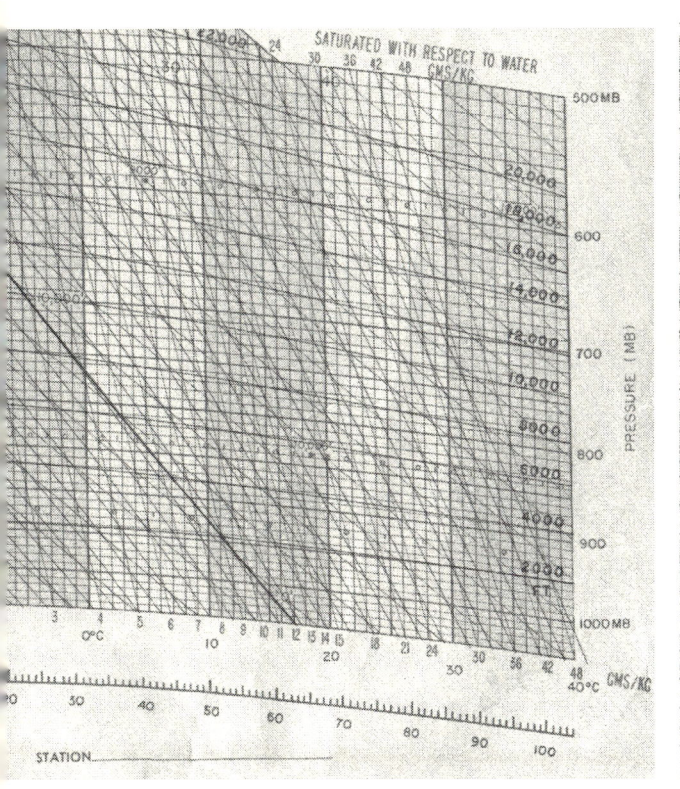

Figure 3-21g. Arowagram. This rare diagram uses a coordinate system of linear temperature (X-axis) and curved pressure (Y-axis), combining the gridlike attributes of the Stüve diagram with the logarithmic, quasi-horizontal pressure scale of the modern tephigram.

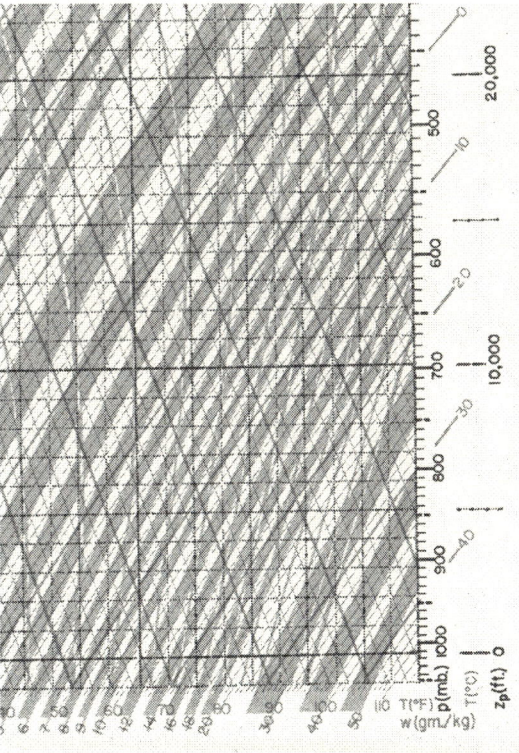

Figure 3-21h. Pastagram. This is one of the most obscure thermodynamic diagrams in weather history. It is similar to the Refsdal diagram with linear entropy (X-axis) and logarithmic pressure (Y-axis). This specimen dates to 1946.

52 • WEATHER FORECASTING HANDBOOK

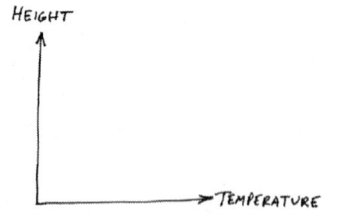

Figure 3-22a. Soundings are not confusing when you keep in mind that they simply allow us to plot a weather reading on a graph of height versus temperature. If the lower atmosphere has a high temperature, you plot a dot at the bottom right. If the upper atmosphere has a low temperature, you plot a dot at the top left. The dots can be connected to show a temperature profile.

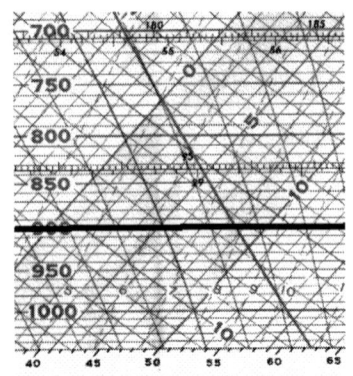

Figure 3-22b. Pressure (height) lines run parallel to the thick line shown.

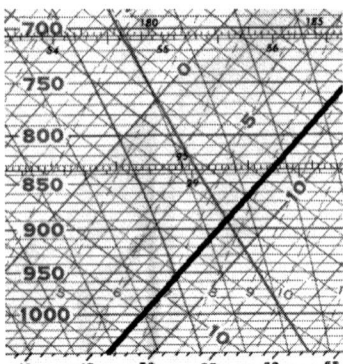

Figure 3-22c. Temperature lines run parallel to the thick line shown. Together, the pressure (height) and temperature line help locate the temperature of a parcel at a given height. Everything else is secondary.

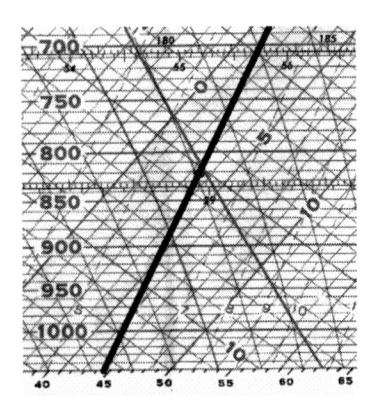

Figure 3-22d. Mixing ratio lines run parallel to the thick line shown. They help identify the mixing ratio or the saturation mixing ratio of an air parcel.

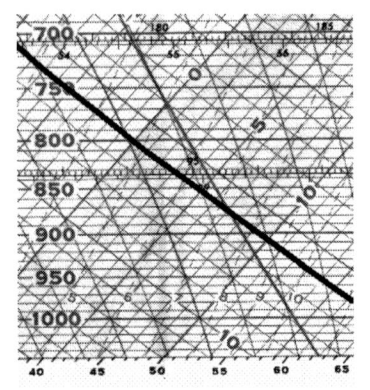

Figure 3-22e. Dry adiabat lines run parallel to the thick line shown. When a parcel rises or sinks without any phase change, it can only do so along a dry adiabat. More about this is covered later in the text.

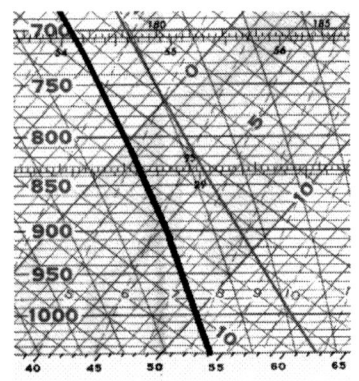

Figure 3-22f. Wet adiabat lines run parallel to the thick line shown. When a parcel rises and it is saturated (condensing), it can only rise along a wet adiabat. Parcels only sink along dry adiabats. More about this is covered later in the text.

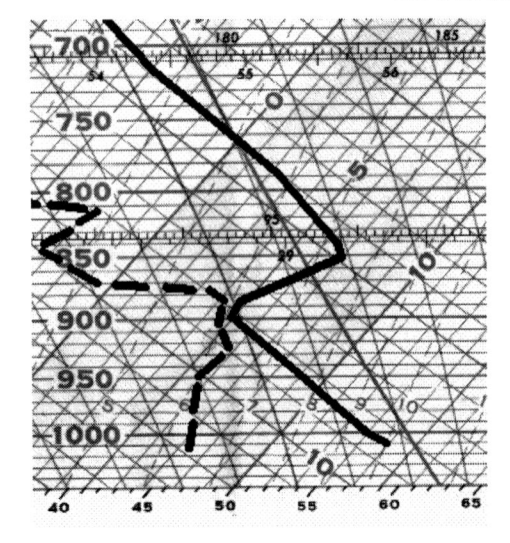

Figure 3-23. An example of a sounding. The temperature profile is indicated by the solid line on the right. It shows a temperature of 14 deg C at the surface, 5 deg C at 900 mb, 7 deg C at 850 mb, and so forth. The dewpoint (moisture) profile is indicated by the dashed line on the left. It shows a dewpoint of 7 deg C at the surface, 5 deg C at 900 mb, -3 deg C at 850 mb, and so forth.

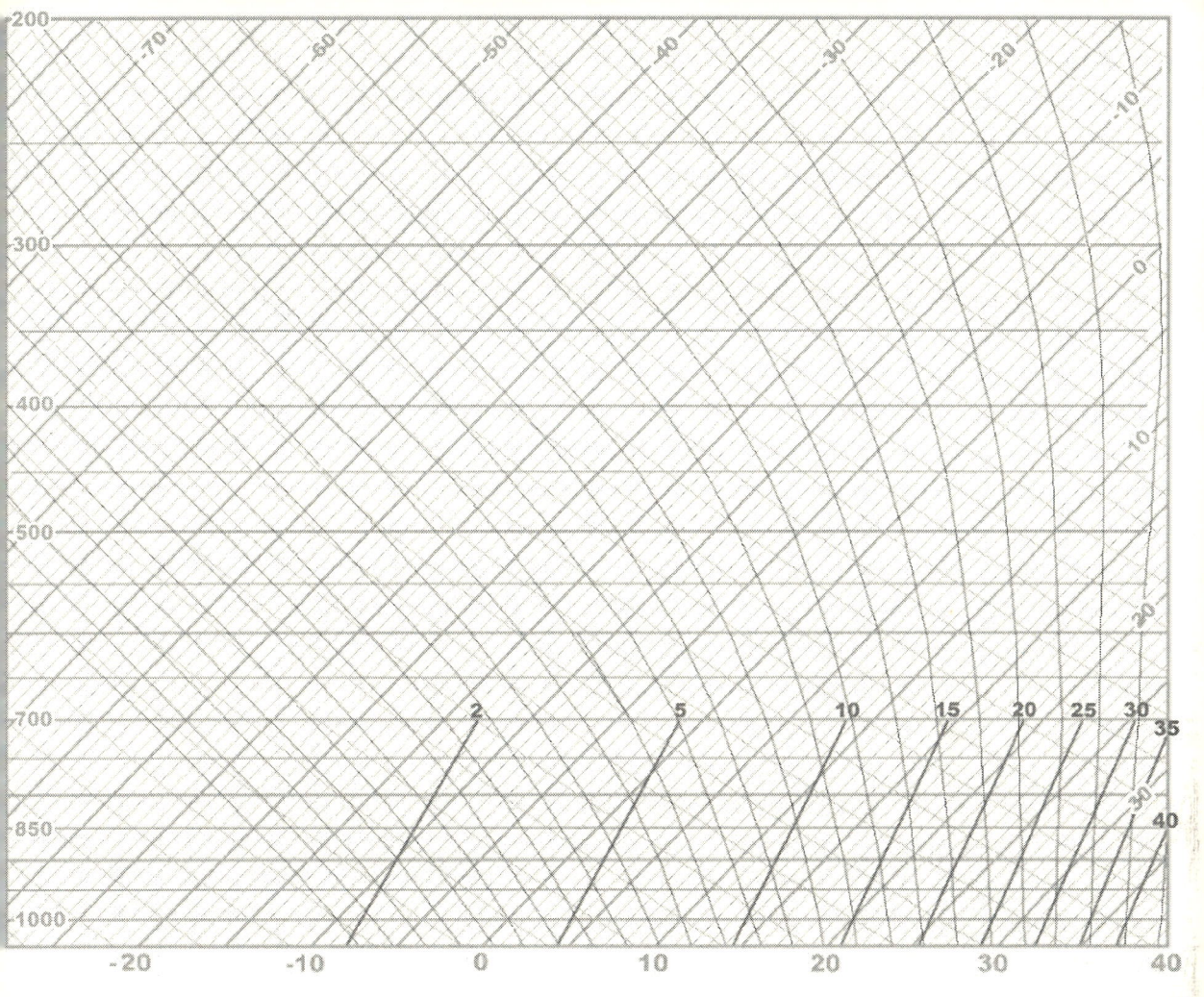

Figure 3-24. Blank Skew-T diagram. It gives us a coordinate system for plotting the vertical temperature profile through the atmosphere. Height increases toward the top of the chart, and temperature increases toward the right side of the chart. This diagram may be freely copied and used.

Figure 3-25. Radiosonde launch by Shannon Key, a National Severe Storms Laboratory research assistant, in Ajo, Arizona. The data acquired from weather balloons yields soundings and hodographs, and are frequently ingested as input into computer model forecasts.

connected in order of increasing height, and the resulting line is known as the hodograph trace.

Vector operations can easily be done on the hodograph, which allow quantities such as shear and helicity to be easily computed. The length of a hodograph trace over a fixed vertical distance (such as 2 to 3 km) is proportional to the shear through that layer.

Techniques for using the hodograph are detailed in the Convective Weather chapter.

3.7. LIGHTNING DETECTION

Unfortunately lightning detection is a forecasting tool that remains closely guarded by "big business" interests, placing it out of the hands of many meteorologists and emergency management officials. The U.S. government has no publically-funded regional lightning network. Standard lightning detection networks are capable of sensing cloud-to-ground lightning ONLY, and can distinguish polarity of those strikes.

Indications. Lightning detection is an excellent tool for determining the energy in a developing synoptic or mesoscale storm system. Furthermore it can alert forecasters to thunderstorm development in remote areas not sampled well by radar. There is a correlation with lightning frequency and heaviest rainfall. The heaviest rainfall tends to occur where the frequency is highest. Over 95% of cloud-to-ground strikes are negative (lowering negative charge to the surface). Positive strikes are more rare, and are more powerful and last longer. Positive strikes tend to highlight areas of deep convection. Concentrated areas of positive strikes are indicative of severe thunderstorms, and may be associated with supercell storms.

REVIEW QUESTIONS

1. Name some common markings on the 500 mb chart.

2. What would have a higher weather radar reflectivity: a thick dense cloud with no precipitation or a thin cloud with light precipitation?

3. Which of the following indicators is the best indicator of possible tornadoes, hail, or high winds? A) a heavy reflectivity core that is near the edge of an echo; B) a large region of

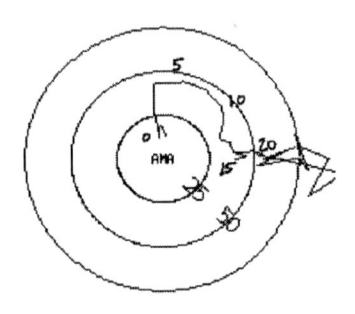

Figure 3-26. Hodograph showing strong turning of the winds in the lowest 2 km (6000 ft) of the atmosphere. Heights are labelled in thousands of feet. Strong rotating storms developed this day, but low moisture precluded tornado formation.

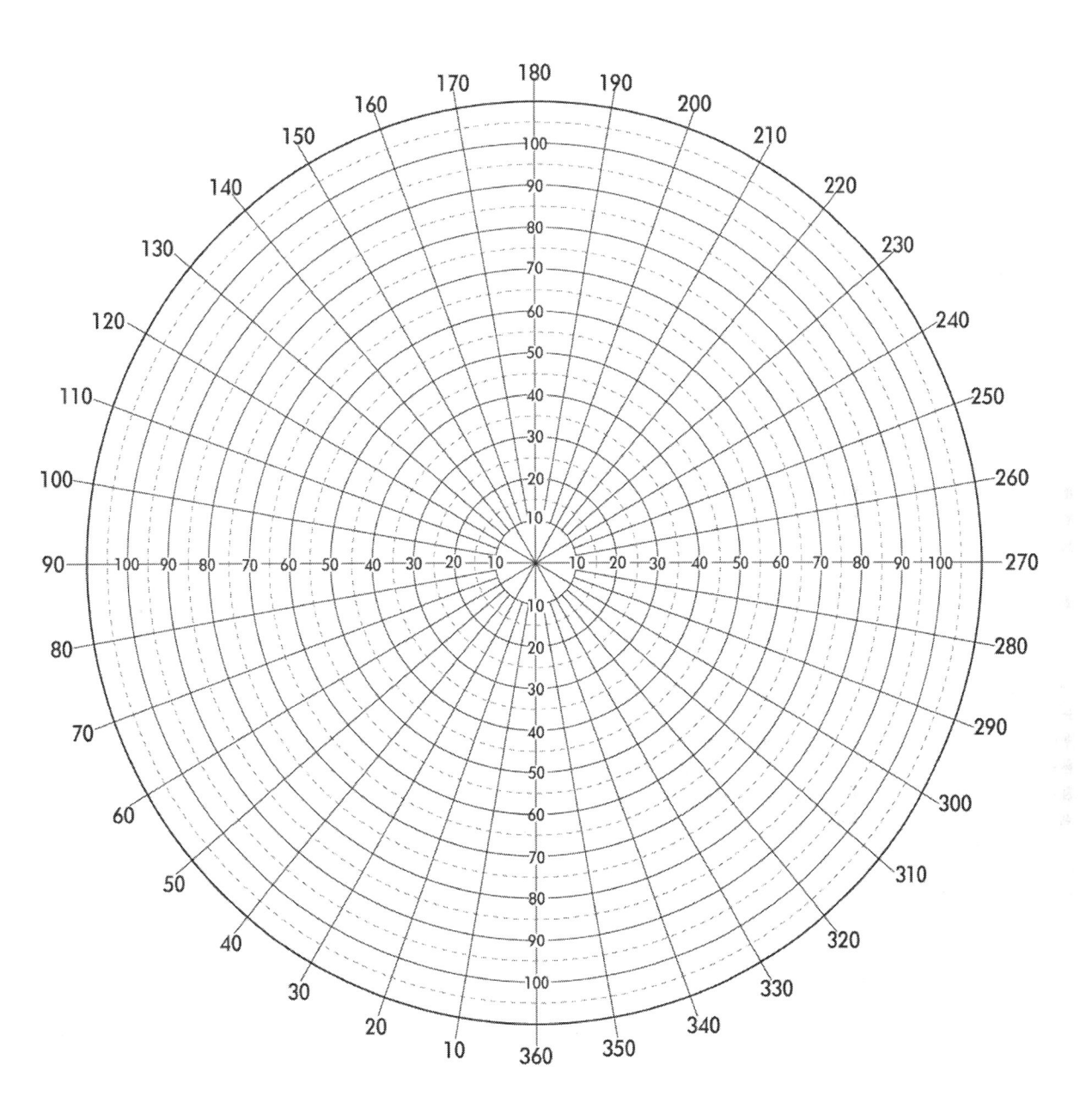

Figure 3-27. Blank hodograph. This page may be copied for personal use.

precipitation spreading downstream; or C) two distinct cells merging.

4. What causes a hook echo pattern?

5. Describe in detail the pattern and orientation of a Kansas tornado signature as would be seen on Doppler radar velocity products.

6. You are forecasting weather in Louisiana and it's 3 a.m. You notice on the infrared satellite imagery that various lakes which are standing out warm on the imagery are starting to "disappear". What is happening and what does it mean?

7. You are looking at water vapor imagery of Florida and note a bright area of moisture. Is this more likely to represent a rich area of low-level moisture or a weak area of upper-level moisture? Why?

8. A hodograph trace shows a round "O" shape centered on the graph origin (its center). What does this imply about the wind speed and wind direction in the atmosphere?

9. Can a hodograph be plotted based on data from Doppler weather radars? Why or why not?

10. Which shows a higher amount of lightning activity on a standard lightning detection network: a thunderstorm with weak cloud-to-ground lightning activity or frequent cloud-to-air lightning activity?

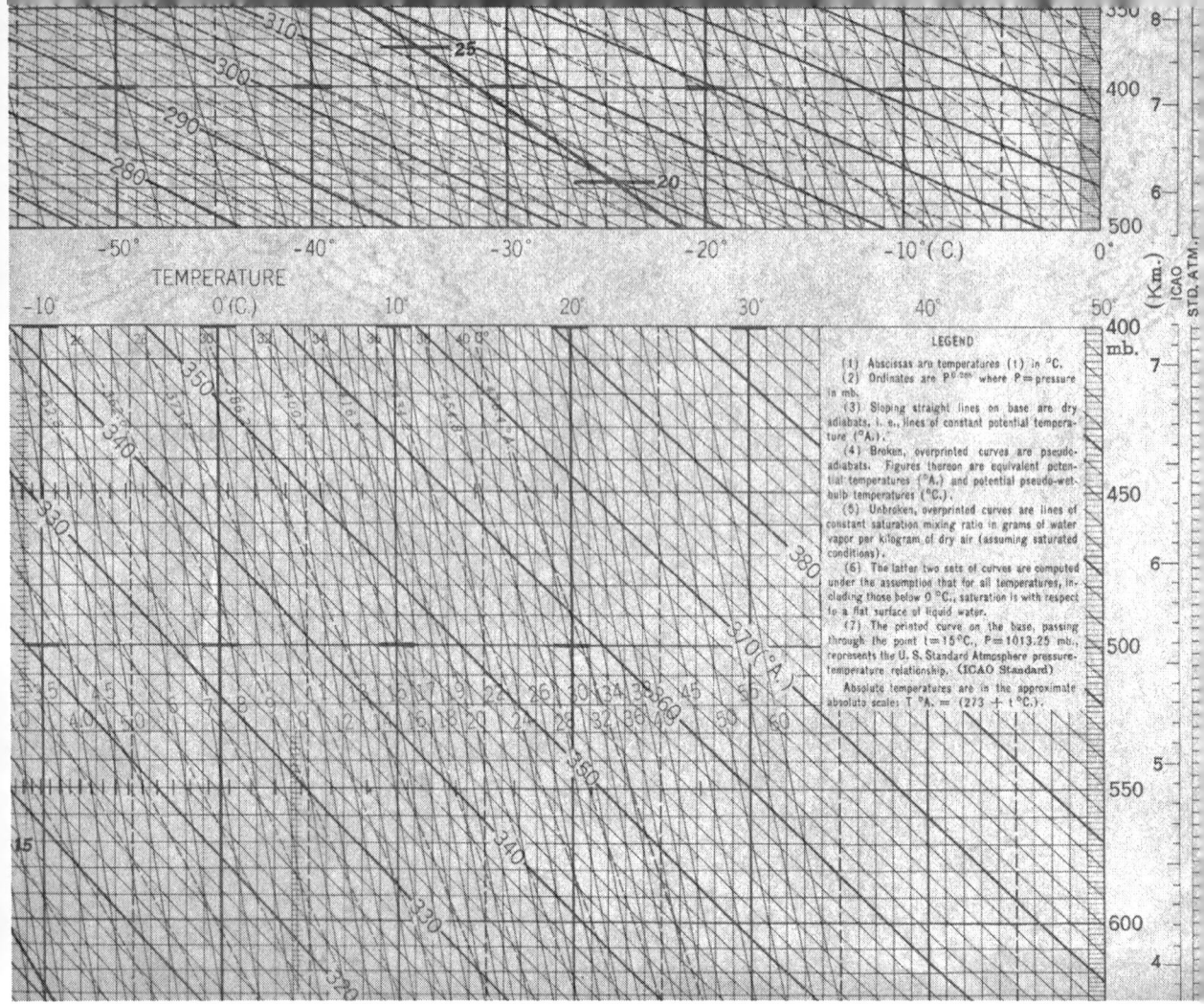

Portion of 1950's National Weather Service pseudo-adiabatic diagram.

4 PHYSICS

Nothing is too wonderful to be true if it is consistent with the laws of nature.

MICHAEL FARADAY

In this chapter we will assemble what we've learned about meteorological fundamentals, observations, and tools, and see how physical processes such as lifting, melting, and changes in pressure gradients can cause changes in the properties of an air mass.

4.1. PHASE CHANGES

Heating and cooling processes are what create weather, and any change in the phase of water (such as ice to water) can release or absorb heat in a parcel. When a phase change occurs over a large area, and it commonly does, the effect on weather can be substantial, ranging from cloud development to evaporational cooling.

<u>4.1.1. Changes requiring heat</u>. Energy is absorbed by water molecules and a parcel is cooled when the water molecules go to a higher energy state. This includes the following phase changes:

4.1.1.1. Melting. Conversion from ice to liquid.

4.1.1.2. Evaporation. Conversion from liquid to vapor.

4.1.1.3. Sublimation. Conversion from ice directly to vapor.

<u>4.1.2. Changes releasing heat</u>. Energy is released by water molecules and a parcel is warmed when the water molecules go to a lower energy state. This includes the following phase changes:

4.1.2.1. Freezing. Conversion from liquid to ice.

4.1.2.2. Condensation. Conversion from vapor to liquid.

4.1.2.3. Deposition. Conversion from vapor to ice (sometimes called sublimation).

4.2. ADIABATIC CHANGES

Adiabatic is used to define a system where no heat or mass is exchanged between a parcel ("sample") of air and its surroundings. Therefore in an adiabatic process we assume that solar heating, conduction, radiation, and so on is not changing the properties of our parcel.

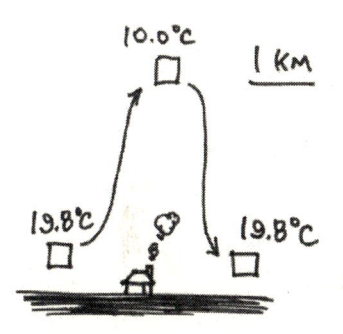

Figure 4-1. A dry parcel's temperature changes at the dry adiabatic lapse rate. This rate is 9.8 C deg per vertical km.

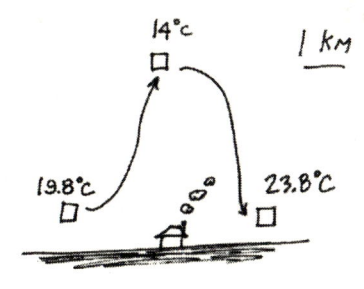

Figure 4-2. A saturated parcel's temperature changes at the wet adiabatic lapse rate, which is roughly 6 C deg per vertical km. Note that a parcel typically sinks at the dry adiabatic lapse rate (9.8 deg per km). As a result, the parcel is warmer when it returns to the surface. This is because wet adiabatic lifting contributes latent heat to the parcel.

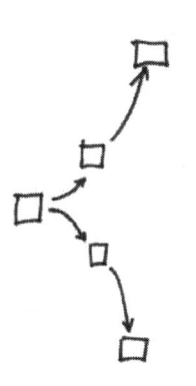

Figure 4-3. A parcel in unstable air tends to rise or sink more quickly.

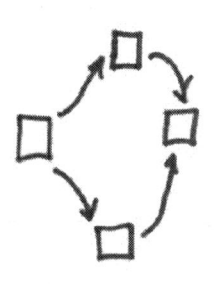

Figure 4-4. A parcel in stable air tends to return to its original level after sinking or rising.

4.2.1. Vertical changes. When a parcel is lifted, it expands and cools. When a parcel sinks, it compresses and warms. This occurs at the rate of about 10 C deg per vertical km. This rate of change is known as the "dry adiabatic lapse rate", and assumes that the parcel is not saturated.

4.2.2. Release of latent heat. If, while being lifted, a parcel cools to its dewpoint temperature (saturates), any further lifting will cause condensation into water droplets. As described earlier, this is a process that releases heat into the parcel, and this contributes as much as 4 C deg of heat per vertical km to the parcel. As a result, the overall cooling rate can change from 10 to 6 C deg per vertical km. This is known as the "wet adiabatic lapse rate." As lifting continues into high levels (above 10 to 20 thousand feet), less and less latent heat is contributed to the parcel due to the lack of moisture and this process becomes less significant.

4.2.3. Absorption of latent heat. Assume that latent heat has been released in a parcel. If all of the droplets remain present and the parcel begins sinking again, warming will occur and the droplets will begin evaporating. This absorbs heat from the parcel, and this continues until all of the droplets have evaporated. In this way, all the latent heat is conserved and theoretically the parcel returns to a given level at its original temperature.

4.2.4. Pseudo-adiabatic processes. When rising and condensation occurs in a parcel, the water droplets usually do not remain in the parcel -- they fall out because of gravity in the form of rain or snow. Since this causes an exchange in mass with the environment, the process is no longer adiabatic but is "pseudo-adiabatic". Latent heat is released, and when the parcel sinks again, little or no absorption of latent heat occurs since the water droplets are no longer there. Therefore the parcel temperature rises at a faster rate than it cooled when it was rising. This explains why air flowing through a mountain range can cause precipitation on the windward side, then descend and warm drastically on the lee side (producing "chinook" winds).

4.3. STABILITY

4.3.1. Tendency for vertical motion. The hydrostatic equation, $\Delta P/\Delta Z = -\rho g$, indicates that if the density term is increased

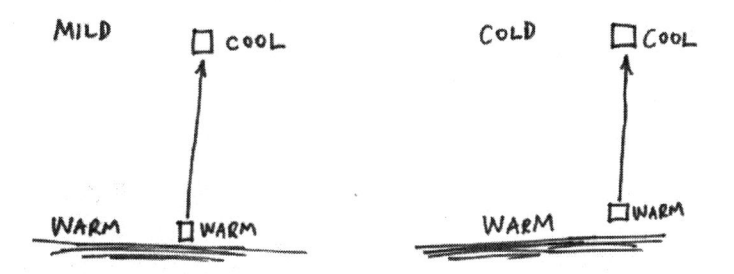

Figure 4-5. When a parcel rises and finds that it is cooler than the air around it, the parcel becomes less buoyant. This type of atmosphere is considered stable.

Figure 4-6. When a parcel rises and finds that it is warmer than the air around it, the parcel becomes more buoyant. This type of atmosphere is considered unstable.

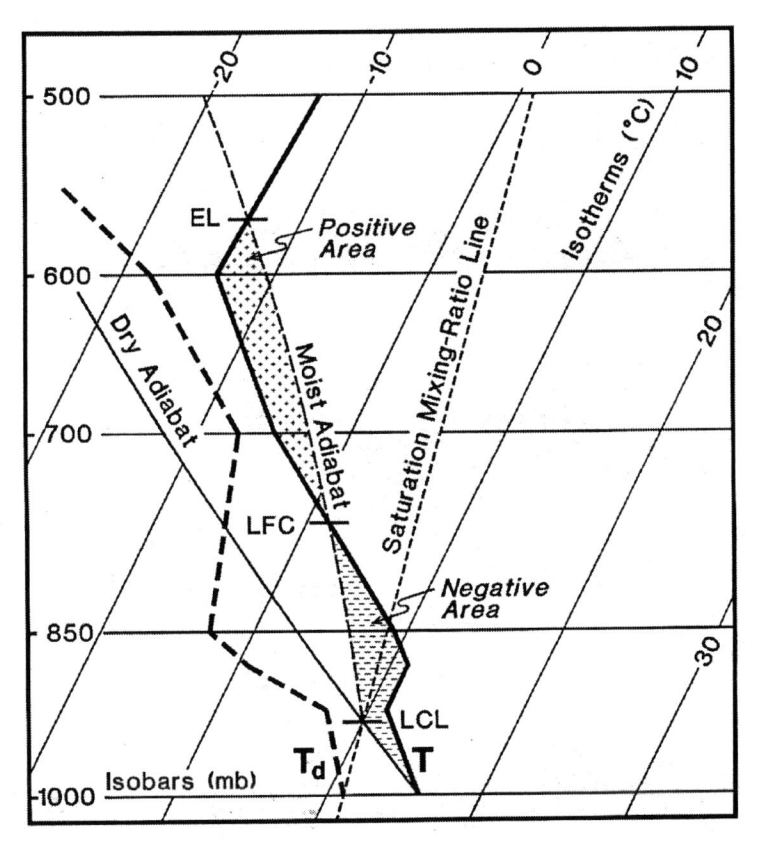

Figure 4-7. Summary of sounding features showing the interrelation of a lifted parcel with its environment. In this example we assume a surface temperature of 15 deg C and a surface dewpoint of 11 deg C. The environmental temperature profile for this day is drawn as a thick solid line (T), and the environmental dewpoint profile is drawn in thick dashed line (T_d). We lift a parcel from the surface, noting its starting mixing ratio. The parcel's temperature always follows the dry adiabat until it meets this mixing ratio line. At this level, condensation occurs, and we consider this level to be the lifted condensation level (LCL). From that point, it follows the moist adiabat upward. This lifted temperature profile is compared against the environmental dewpoint profile. At levels where the lifted parcel is cooler than the environment, a negative energy area is indicated (the parcel is cooler than the surrounding air, thus it gains a sinking tendency). At levels where the lifted parcel is warmer than the environment, a positive energy area is indicated (the parcel is warmer than the surrounding air, thus it gains buoyant tendencies). Wherever the lifted parcel rises into an area of large positive energy, this level is called the level of free convection (LFC). The parcel rises until the surrounding temperature becomes warmer (typically in the stratosphere); the level at which this crossover occurs is called the equilibrium level (EL). The parcel will usually continue rising through its own momentum, and will sink back under the EL with time.

62 • WEATHER FORECASTING HANDBOOK

enough, the parcel will sink. If the density term is decreased enough, the parcel will rise.

4.3.1.1. Rising motion. If the virtual temperature (T_v) of the parcel is greater than that of its environment, the weight of the parcel will be less than the weight of the displaced atmosphere, and the parcel will accelerate upward. For most practical purposes we can use regular temperature in place of virtual temperature.

4.3.1.2. Sinking motion (subsidence). If the virtual temperature (T_v) of the parcel is less than that of its environment, the weight of the parcel will be greater than the weight of the displaced atmosphere, and the parcel will accelerate downward. For most practical purposes we can use regular temperature in place of virtual temperature.

4.3.2. Lapse rate. Lapse rate is the decrease in temperature with height in a given layer (often expressed in deg C per km). It can be measured directly from the sounding or raw data, or can be "eyeballed" on the sounding by evaluating the slope of the temperature trace in the layer. A trace that leans strongly to the left with height cools rapidly with height and is said to have a steep lapse rate, or a "steep slope". A trace that is vertical or leans to the right with height is said to have a weak lapse rate, or a "weak slope".

4.3.3. Stability of the atmosphere. When a parcel in the environment is forced upward or downward by some external force, the lapse rate determines whether the parcel returns to its original level or continues accelerating.

4.3.3.1. Stability. Assume the environment has a low lapse rate, and a parcel temperature is the same as the environment at a given level. If the parcel rises, it cools more rapidly than the environment cools with increasing height, which increases its tendency to sink. If the parcel sinks, it warms more rapidly than the environment warms with decreasing height, which increases its tendency to rise. Therefore, the parcel tends to return to its original level.

<u>Summary of a stable layer:</u>
* Slope: The lapse rate is less steep than the wet adiabat.
* Lapse rate: Less than about 6 C deg per km (sometimes it even warms with height).
* When a parcel is forced upward: It cools at faster rate than the environment, so it becomes heavy and returns to its original level.

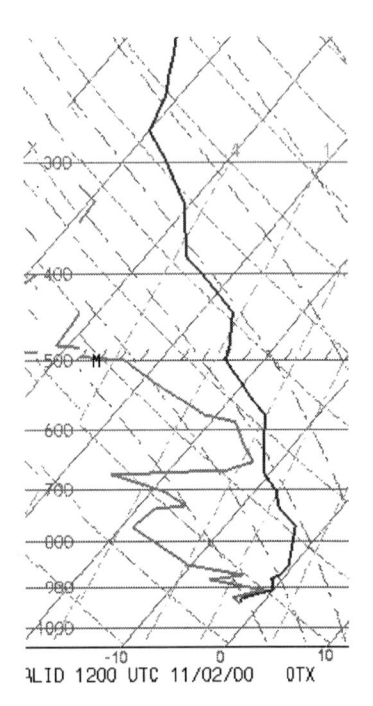

Figure 4-8. The plot of temperature on this sounding (right line) shows that the temperature remains fairly steady up to 850 mb (5000 ft). Since the slope of this section is less than that of the wet adiabats (i.e. it leans to the right with height), the layer is considered stable. Above that level, there are several layers where the slope of the temperature line is greater than the wet adiabat but less than the dry adiabat, such as the layer between 800 and 700 mb. These sections are considered conditionally unstable, which means that a moist parcel starting within such layers can continue accelerating if they are forced to rise or sink.

* When a parcel is forced downward: It warms at a faster rate than the environment, so it becomes buoyant and returns to its original level.

4.3.3.2. *Conditional instability*. If the environment is conditionally unstable (the layer's lapse rate exceeds the wet adiabatic rate but not the dry adiabatic rate), the effect on the parcel depends entirely on whether it is saturated. If it is dry, it will always tend to return to its original level since it becomes too cold when lifted and too warm when sunk. However if it is saturated, any upward motion (turbulent flow, etc) will cause latent heat to be released in the parcel, making it rise at the wet adiabatic rate (along the wet adiabat). It becomes warmer than the surrounding air accelerating its motion. In other words, it is buoyant. This is the scenario that occurs in most precipitation events. However, a saturated parcel that sinks will dry out, warm at the dry adiabatic rate, and return to its original level.

<u>Summary of a conditionally unstable layer:</u>
* Slope: The lapse rate is steeper than the wet adiabat but not as steep as the dry adiabat.
* Lapse rate: Greater than about 6 C deg per km but less than about 10 C deg per km.
* When a parcel is forced upward: If it is dry, it cools at faster rate than the environment, so it becomes heavy and returns to its original level. If it is saturated, it cools at a slower rate than the environment, so it becomes more buoyant and accelerates upward.
* When a parcel is forced downward: It warms at a faster rate than the environment, so it becomes buoyant and returns to its original level.

4.3.3.3. *Absolute instability*. Except near the ground on sunny days, absolute instability is very rare since it often causes immediate effects that restore equilibrium in the layer. Let's see how this works. Assume the environment is absolutely unstable (the layer's lapse rate exceeds the dry adiabatic rate), and a parcel temperature is the same as the environment at a given level. If the parcel rises, it cools less rapidly than the environment cools with increasing height, which further increases its tendency to rise. If the parcel sinks, it warms less rapidly than the environment warms with decreasing height, which further increases its tendency to sink. Therefore, any force that acts vertically on the parcel (turbulent flow, etc) causes it to accelerate in the direction it is displaced.

<u>Summary of a conditionally unstable layer:</u>
* Slope: The lapse rate is steeper than the dry adiabat.
* Lapse rate: Greater than about 10 C deg per km.

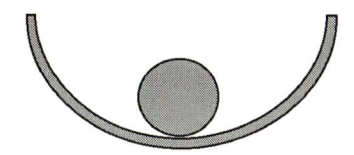

Figure 4-9a. STABILITY. Here we see a ball and a bowl. The ball is at a state of rest, and any force will cause it to return back to its starting point. In a similar manner, a parcel that rests in a stable layer will always return to its original level when an upward or downward force is applied.

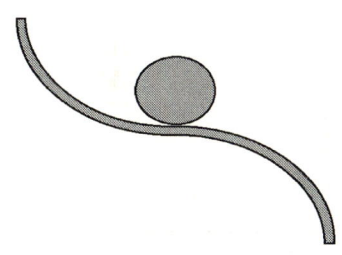

Figure 4-9b. CONDITIONAL INSTABILITY. When a layer is conditionally unstable, a parcel will always return to its starting point when downward force is applied (i.e. pushing the ball above to the left). This will also occur if the parcel is <u>dry</u> and an upward force is applied. However if the parcel is <u>saturated</u>, an upward force will cause acceleration upward (i.e. pushing the ball above to the right).

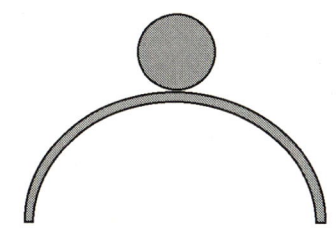

Figure 4-9c. ABSOLUTELY UNSTABLE. If a parcel is forced either up or down, it will continue accelerating in that direction.

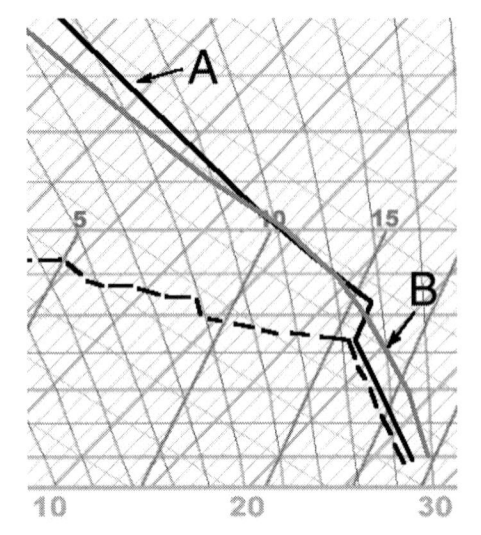

Figure 4-10. Example of potential instability (see text). The entire layer represented by the temperature trace (dark solid line, A) is lifted by 100 mb. The result is the solid gray line (B). It can be seen that the lower layers have warmed (due to release of latent heat), while the upper layers have cooled (due to adiabatic cooling). The result is a temperature profile that has a steeper lapse rate and is trending towards becoming absolutely unstable. Note that the mid-level inversion has been eroded by the lift.

* When a parcel is forced upward: It cools slower than the environment, so it becomes more buoyant and continues accelerating upward.
* When a parcel is forced downward: It warms at a slower rate than the environment, so it becomes heavy and continues accelerating downward.

4.3.4. Potential instability. This is a different type of instability altogether, which involves deep layers of air. Rather than lifting parcels, we are forcing the lifting an entire layer (i.e. every parcel in the layer). This is frequently what occurs with dynamic sources of lift (upper level divergence, surface convergence, etc). The result is that saturated levels, particularly in the low levels, tend to show warming, while dry levels tend to show cooling. The net effect can steepen the overall lapse rates, making the atmosphere more unstable and eroding inversions that hinder convection. Such a situation frequently occurs in the Great Plains in the springtime.

4.4. ATMOSPHERIC FORCES

4.4.1. Coriolis Force. The Coriolis force is an apparent force that acts on objects moving across Earth's sphere. It occurs because of the change in angular velocity with latitude when viewed within an Earth-based rotating coordinate system. The Coriolis force produces a rightward deflection on any moving air parcel in the northern hemisphere (leftward in the southern hemisphere), with a magnitude directly proportional to its velocity.

4.4.1.1. Definition. The Coriolis force equals fv. The term f is $2\Omega \sin(\phi)$, where Ω is the Earth's angular velocity (a constant), and ϕ is the latitude. Therefore f is zero at the Equator and increases with latitude. The term v is velocity of the parcel's motion.

4.4.1.2. Implications. The above definition shows that the Coriolis force is zero when the parcel is stationary. When the parcel is in motion, Coriolis force is directly proportional to the velocity of the air parcel. The Coriolis force also increases in strength as one moves away from the Equator.

4.4.1.3. Is it a force or an effect? This is an interesting argument. In an initial coordinate system there is not a Coriolis force. However all motions are measured in the Earth's rotating coordinate system, where the Coriolis effect does act as a force. Although physically the Coriolis force cannot do work, this is not necessarily a prerequisite for it to be a force.

4.4.2. Pressure Gradient. This equals $\Delta P/\Delta n$ (change of pressure per change of distance), and is always measured perpendicular to the isobars.

4.4.3. Pressure Gradient Force. This is a force that acts on a parcel of air, resulting from the pressure gradient surrounding a parcel. It equals $-(1/\rho) \cdot \Delta P/\Delta n$. In short, it acts to move parcels toward lower pressure.

4.4.4. Contour Gradient. This equals $\Delta z/\Delta n$ (change of height per change of distance), and is always measured perpendicular to the contours. This is similar to the pressure gradient but since we look at contours (lines of constant height) on upper air charts, this is another way of looking at pressure differences aloft.

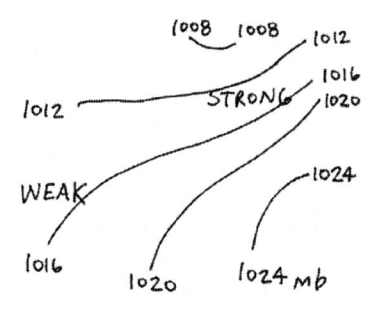

Figure 4-11. A weak pressure gradient occurs where isobars are far apart.

4.4.5. Contour Gradient Force. This is a force that acts on a parcel of air, resulting from the contour gradient surrounding a parcel. It equals $-(1/\rho) \cdot \Delta z/\Delta n$. In short, it acts to move parcels toward lower heights.

4.4.6. Centrifugal Force. The apparent force that deflects particles (winds) away from the center of rotation. It is always directed outward from the axis of rotation and is *perpendicular* to the direction of v (velocity). We use it to describe the balance of forces as winds flow around highs, around lows, and anywhere that the flow is not a straight line.

4.4.7. Frictional Force. Friction is a force that opposes motion. It always acts *opposite* to the direction of v (velocity). If friction increases, velocity decreases.

4.5. WINDS

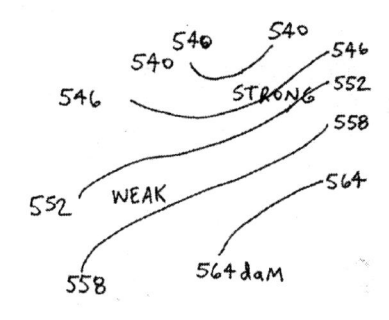

Figure 4-12. A weak contour gradient occurs where contours are far apart.

4.5.1. Geostrophic wind. Geostrophic wind is an imaginary wind that would result if there was an exact balance between the Coriolis force and the pressure gradient force. It can easily be calculated by analyzing only the existing pressure or contour gradient and the latitude at each point. For a geostrophic wind to happen, the flow would have to be straight-line (no centrifugal force) and there could not be any friction (naturally winds are nearly geostrophic as you go up in height).

4.5.2. Supergeostrophic wind. If the actual wind is flowing faster than the geostrophic wind, the Coriolis force becomes dominant and forces the air to flow towards higher heights (where it loses kinetic energy) until the velocity slows down and thus the Coriolis force subsides. This usually happens around the exit (east) region of jet streaks.

4.5.3. Subgeostrophic wind. If the actual wind is flowing slower than the geostrophic wind, the pressure gradient force becomes dominant and forces the air to flow toward lower heights (where it gains kinetic energy) until the velocity speeds up and the Coriolis force becomes more dominant. This usually happens around the entrance (west) region of jet streaks.

4.5.4 Gradient wind. This is the wind that would result of there was a balance between the pressure gradient force, the Coriolis force, and centrifugal force. It assumes no friction.

4.5.4.1. Anticyclonic (supergradient) wind. The Coriolis force (acting to the right in the northern hemisphere) balances against a combination of the pressure gradient force and centrifugal force (acting to the left). Since the centrifugal force will help the parcel accelerate into lower pressures (increasing its kinetic energy), it speeds up until the Coriolis force strengthens. Therefore the wind is stronger in anticyclonic flow than what the geostrophic wind would indicate.

4.5.4.2. Cyclonic (subgradient) wind. The Coriolis force and centrifugal force combine to balance against the pressure gradient force. Since the centrifugal force helps the parcel move away from lower heights (reducing its kinetic energy), it slows down until the Coriolis force weakens. Therefore the wind is weaker in cyclonic flow than what the geostrophic wind would indicate.

4.5.5. Cyclostrophic wind. When Coriolis force is negligable (such as near the Equator) where cyclonic gradient winds exist, centrifugal force becomes the main force that balances the pressure gradient force. The winds in hurricanes and typhoons are often near cyclostrophic balance, as are tornadoes and dust devils. Either clockwise or counterclockwise flow is possible.

4.5.6. Frictional effects. Friction always opposes the velocity, which helps negate the effects of the Coriolis force. Therefore

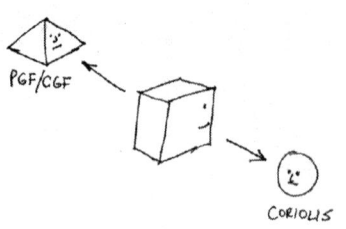

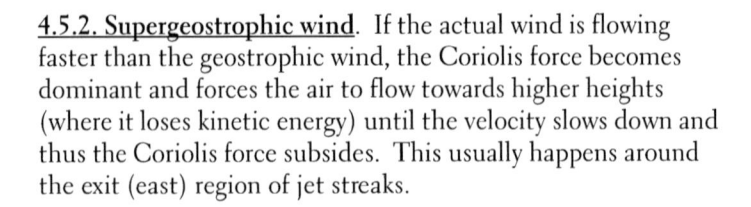

Figure 4-13. A parcel is in geostrophic balance when the pressure gradient force or contour gradient force (its tendency to flow toward lower pressure) is balanced out by the Coriolis force (which is proportional to velocity and latitude).

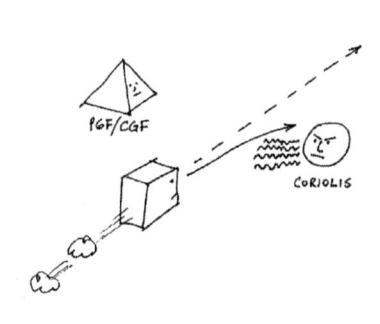

Figure 4-14. Supergeostrophic flow occurs when a parcel is moving faster than the speed at which it would have geostrophic balance. The higher speed causes an increase in the Coriolis force, and the parcel bends toward the right. This commonly occurs as upper-level parcels leave areas of high winds.

when friction is added, the pressure gradient force becomes dominant, helping the parcel turn toward lower pressure. This is why winds flow more directly into low pressure over land than over oceans. It also explains why winds tend to back and slow down as one moves downward toward the Earth's surface.

4.5.6.1. Boundary layer. The boundary layer is that part of the atmosphere where friction from the Earth's surface affects air motion. The top of the boundary layer (called the gradient level) is is usually 2300 ft (700 m) but is lower over the oceans and higher over mountainous terrain. Higher wind speed and higher instability tend to increase the gradient level due to increased turbulence and mixing within the boundary layer.

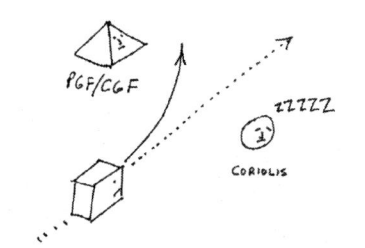

Figure 4-15. Subgeostrophic flow occurs when a parcel is moving slower than the speed at which it would have geostrophic balance. The slower speed minimizes the effect of the Coriolis force, and the parcel bends toward the left toward lower pressures. This commonly occurs as upper-level parcels enter areas of high winds.

REVIEW QUESTIONS

1. What type of phase change requires thermal energy to free the bonds of water molecules?

2. The tallest buildings in the world, the Petronas Towers, are 452 meters tall. In an elevator that starts at street level and reaches the top, is the interior warmer on a dry day or a wet rainy day? What is the total temperature change in the elevator in deg C? This assumes no external influences like air conditioning, evaporation, and so forth.

3. In the last question, assume the elevator goes to the top without opening, then returns to street level. Any condensation that occurs is instantly absorbed by the carpet inside. Will the air inside be warmer on a dry day or on a wet rainy day? By how much will it have warmed on the trip from the top to the bottom?

4. In the last question, if there is no carpet in the elevator and any condensation evaporates immediately on the way down, what is the elevator air's final temperature at street level compared to the last time it was at street level?

5. If the 700-500 mb layer in the atmosphere shows that there is a lapse rate of 7 deg C per km, is the layer stable, conditionally unstable, or absolutely unstable?

6. Is the Coriolis force stronger at the Equator or at the North Pole?

7. In what type of wind flow does air gain kinetic energy?

8. Is a parcel of air embedded in a jet stream more likely to speed up or slow down as it rounds the curve of a ridge, turning anticyclonically? Assume that the contour gradient force is constant. What kind of wind is this called?

9. Why is it uncommon for strong, persistent surface high pressure areas to be found on the Equator?

10. Do low pressure areas fill more quickly over ocean or over land? Why?

Mammatus and cumuliform clouds near Muskogee, Oklahoma in 1998. *(Tim Vasquez)*

5 FRONTS AND JETS

I've lived in good climate, and it bores the hell out of me. I like weather rather than climate.

JOHN STEINBECK
Travels With Charley, 1962

Finally our discussions bring us to the familiar weather systems that affect the temperate and polar regions. Frontal systems and upper-level jets (ribbons of strong winds) are symbiotic entities that have a significant influence on one another. They are created by the close proximity of air masses that have different temperature contrasts. This contrast sharpens pressure gradients aloft, which in turn create complex flows aloft that can cause surface pressures to fall. This in turn brings the air masses closer together, stimulating a chain reaction that we will learn about in the chapters ahead.

5.1. AIR MASSES

Air masses are typically classified by their source region moisture character (continental, c; or maritime, m) and their source region temperature character (arctic, A; polar, P; tropical, T; or equatorial, E). They are combined as shown in the entries below, and these abbreviations should be marked on surface maps to help note the location of these air masses. In addition to this, meteorologists occasionally suffix this designation with a temperature indicator of whether the air mass is colder or warmer than the surface that it is currently passing over (w, warm; or k, cold). For example, a mild maritime polar air mass which is passing over a frozen expanse of ocean is referred to as mPw, or maritime polar warm. Air masses that are relatively cold (k) will tend to warm from the bottom up, increasing its lapse rate and making it more unstable; such air masses are associated with gusty, turbulent winds that disperse haze and smoke. Air masses that are relatively warm (w) will tend to cool from the bottom up, decreasing its lapse rate and making it more stable; such air masses have stable or inversion conditions with stratiform cloud tendencies and visibilities restricted due to haze, fog, or smoke. These third-level indicators provide a convenient description of the character of the air mass in question and should be used whenever possible.

5.1.1. Continental polar (cP). The cP air mass forms due to radiational cooling of a layer of air over cold or frozen terrain. Its most common source in North America is northwestern Canada. The cP air mass forms the majority of cold air outbreaks into the United States.

5.1.2. Continental tropical (cT). The cT air mass usually forms from insolation in dry terrain. It is dry throughout its depth, and is very warm at the surface. The dryline, a surface feature found in the Great Plains, separates maritime tropical air from cT air.

In the years 860 and 1234, goods were transported from Venice to the opposite [Yugoslavian] coast over the frozen Adriatic . . . In 1305 and 1364 all the rivers were frozen in France, and on the Rhone the ice was in some parts 15 feet thick. In 1709, the sea at Cette and Marseilles was covered to a great distance with ice; most of the fruit trees were destroyed by the cold, and daily the bodies of persons who had been frozen to death were found on the roads.

DR G. HARTWIG,
"The Aerial World," 1886

5.1.3. Maritime polar (mP). Maritime polar air forms when an air mass loses its heat over cold ocean surfaces while gaining moisture from evaporation. This makes it cold and moist in its lower levels. Most mP invasions into the United States are the result of Pacific weather systems.

5.1.4. Maritime tropical (mT). Maritime tropical air forms when an air mass gains heat over warm ocean surfaces while gaining moisture from evaporation. This makes it a warm and very moist air mass. Maritime tropical air contains abundant and rich moisture, but since the ocean, not the air mass, gets the most energy from solar heating, convective instability is usually weak in the air mass and is released only through localized heating, surface convergence, or upper-level divergence. However, where mT air wanders onshore, the air mass can gain a lot more heat, and convection becomes a strong possibility. This convection is typically diurnal, occurring around the time of maximum heating.

5.1.5. Arctic (A). There are very minor differences between continental polar (cP) air and arctic (A) air. Although some classifications designate air in the Arctic basin dominated by high pressure as arctic air, the differences are limited mainly to the middle and upper troposphere, where the temperatures are lower in the arctic air mass.

5.1.6. Equatorial (E). Equatorial is usually considered to be an air mass near the equator which is cooler than the air masses further north and south. It is assumed that those air masses north and south of it are subsiding into the subtropical high and have been warmed by adiabatic compression. Typical temperatures in the equatorial air mass are in the 70s and 80s.

5.2. FRONTAL CONCEPTS

A front is a boundary between two different air masses that have different temperatures. The frontal zone is the actual region of temperature contrast between two air masses, and is characterized by packing of isotherms or thickness lines.

5.2.1. Frontal location. The front always exists on the warm side of the frontal zone. In other words, the spot where the warm air begins showing a transition to cooler air marks the location of

---

**HUMOR BREAK —
Maritime Ghost (mG) airmass**

When you watch horror movies, you'll often notice that an electrical thunderstorm is present when evil things are happening. This is a very common meteorological effect around haunted houses and raises many intriguing questions. Anyone who has read the Amityville Horror or owns a haunted house knows that when you go into a room containing an evil presence, the room is usually drafty and cold. This is because the presence of the entity absorbs heat from the room. Eventually, the haunted house grows somewhat cooler, forming an outflow of cold "maritime ghost" air. Convergence along this boundary produces additional thunderstorm development. An exceptionally evil entity will draw larger amounts of heat, causing a more extensive outflow and sharper boundary layer convergence that reinforces storm development (Nightmare on Elm Street, Friday the 13th, et al). On the other hand, the availability of moisture can create high theta-e values and contribute to violent thunderstorms, as is often the case in Transylvania with moisture advection from the Black Sea (where orographic lift up the sides of mountains and castles aids in storm formation).

TIM VASQUEZ
"So, Ya Like Weather?," 1988

the front. Fronts almost always lie in pressure troughs and show cyclonic curvature of winds across it.

5.2.2. Frontal surfaces. Fronts do extend upward into the atmosphere, sloping back toward the cooler air, and this upward extension of the front is referred to as the frontal surface. The surface front exists at the intersection of the frontal surface and the ground surface. An upper front exists at the intersection of the frontal surface and a given upper-level "surface" (such as the 850 mb level).

5.2.3. Frontal inversion. The frontal surface slopes back over cold air, so if a balloon is released in the cold air north of a front, it will soon pass into the warm air mass that exists aloft. As the balloon passes through the frontal zone it will show a reduction in the lapse rate or even warming with height. This is seen as an inversion on soundings, and it is called the frontal inversion.

Figure 5-1. Fronts are always located on the warm side of a temperature gradient.

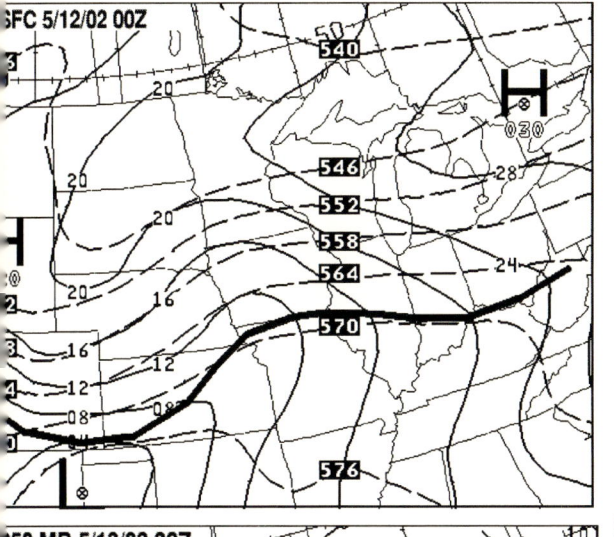

Figure 5-2. Frontal slope. Note that on the surface chart (left) a warm front extends from southern Ohio to southern Iowa to the southeastern tip of Nebraska, where there is a low pressure area (smoothed out by the computer analysis). It then extends southwest as a cold front into the southwestern tip of Kansas. On the 850 mb chart (below left), the warm front is about 200 miles north of this position, and on the 700 mb chart is yet another 200 miles north. This shows how, for example, the frontal surface at Chicago is at the 850 mb level, suggesting the cold air is roughly 5000 ft deep. The cold front also has similar slope at the three different levels, but note how the surface - 850 mb position converges in southwestern Kansas. This is because the elevation of the surface terrain is 4000 ft, close to the 850 mb level, so in this area the surface chart is showing some of the same patterns as the 850 mb chart.

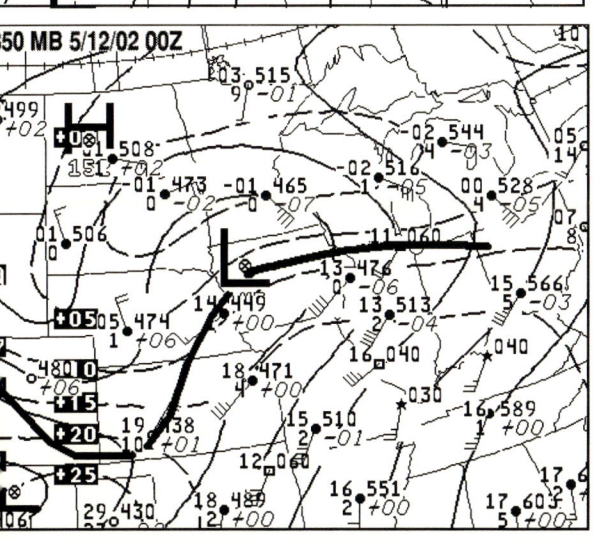

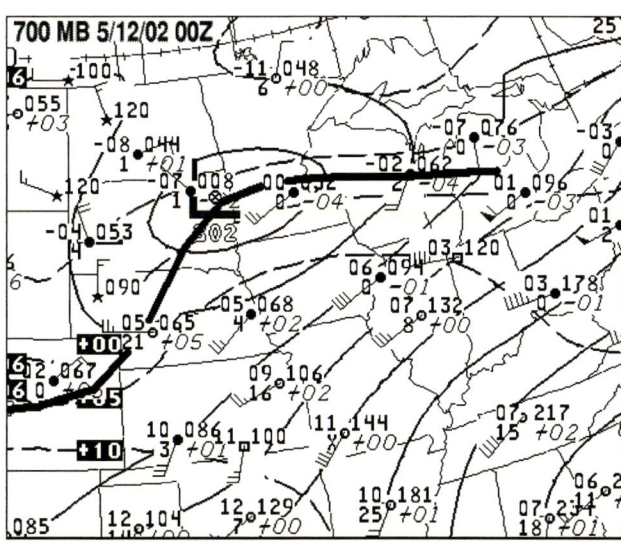

> **Wind shifts and fronts**
>
> Lines of wind shift with no proximity to bands of strong temperature contrast, moreover, appear relatively often on surface charts. The origins of such lines are not typically well-known, and they may arise from more than one source. The widespread practice of analyzing fronts along such wind shift lines is not appropriate. Such lines, including prefrontal wind shifts, should be denoted in some manner to distinguish them from true fronts and other surface boundaries.
>
> CHUCK DOSWELL,
> "A Case for Detailed Surface Analysis," 1995

> **Blue norther**
>
> The temperature along the Texas Rio Grande had hit 100 deg F on the afternoon of February 3, 1899. However, as so often happens in late winter and early spring, the weather scene in Texas underwent a sudden turnabout. With nothing but strands of barbed wire separating the plains and prairies of Texas from the frozen tundra of the north polar region, a ponderous mass of glacial air poured through the state, forcing temperatures to unparalleled depths. By the time the Arctic air settled over Texas on the morning of Lincoln's birthday, the scales on some thermometers were nearly inadequate to gauge the intensity of the severe cold. Readings bottomed out below 0 deg F in virtually all of the northern two-thirds of Texas. The chill lowered the temperature to minus 23 deg F at Tulia in the southern Texas panhandle.
>
> GEORGE W. BOMAR
> "Texas Weather," 1983

The frontal surface exists at the top of the frontal inversion. The dewpoint curve on the sounding usually shows an increase in dewpoint with height through the inversion, as opposed to radiational inversions where the dewpoint usually decreases through the inversion.

5.2.4. Frontal movement. The front typically moves at a speed equalling the component of the wind flow across the front.

5.2.5. Frontal slope. This term refers to the degree of slope of the frontal surface over the cold air. It is usually measured in terms of number of miles of run per mile of rise. For example, a frontal surface with 1:50 (1 mile of rise per 50 miles of run) slope is considered to be steep, while a frontal surface with a 1:300 slope is considered shallow. A steep slope indicates that the lift could be strong if the wind flow forces the air to ascend the frontal surface. Frontal slope in the boundary layer is usually much steeper than that in the middle and upper atmosphere.

5.2.6. Sectors. The warm air mass along a front is usually referred to as the "warm sector", while the cold air mass along a front is referred to as the "cold sector".

5.2.7. Frontogenesis. This term refers to the creation of a new front or the intensification of an existing front. This is reflected by an increase in the temperature gradient, an increase in thickness gradient, or strengthening of the frontal inversion. Frontogenesis is supported by low-level convergence in the wind field along its length, which tends to bring thermal contrasts together and makes them more intense. Frontogenesis can also be supported by diabatic processes (such as surface heating in the warm sector or surface cooling in the cold sector).

5.2.8. Frontolysis. This term refers to the dissipation of a front or the decrease in intensity of an existing front. This is reflected by a decrease in temperature gradient, a decrease in thickness gradient, or decrease in the strength of the frontal inversion. Frontolysis is supported by low-level divergence in the wind field along its length, which tends to push apart existing thermal contrasts and make them less intense. Low-level divergence may be supported by upper-level dynamics. Frontolysis can also be supported by diabatic processes (such as surface heating in the cold sector or surface cooling in the warm sector).

5.3. COLD FRONT

A cold front represents a front where a cold air mass is replacing a warmer air mass. Pressure tendencies tend to show a marked rise after the front passes.

5.3.1. Active cold front (anafront). This is a term used to describe a front where the warm air mass is being forced to ascend the frontal surface. It is associated with steep slope, uniform packing of thickness contours along its gradient, and a polar front jet axis which lies parallel to the surface front. Precipitation and clouds tend to exist behind the cold front.

5.3.2. Inactive cold front (katafront). This term describes a front where the warm air mass is being forced to descend the frontal surface. It is associated with a shallow slope, thickness contours which are not generally parallel to the front, and a polar front jet axis which is not parallel to the front. Precipitation, if any, tends to exist in convective lines ahead of the cold front. Skies are generally clear behind the front.

5.4. WARM FRONT

A warm front is a front in which a warm air mass is replacing a cold air mass. Pressure tendencies tend to stop falling or rise

---

How many kinds of storms are there, how do the depressions of the air-column originate? For of course these do not make their appearance independently, but are caused by former air currents of the width of the polar and equatorial currents. To solve this problem a knowledge of the distribution of winds and the pressure over the whole hemisphere and of the upper currents seems to be necessary. Nothing is so interesting in this regard as this question, but who will before long give the answer? There is besides also necessary the consideration of the direction of the mountain ranges, the outline of the continent, etc. The whole problem is so complicated and the observations so insufficient that we may congratulate ourselves if we contribute something to make the solution easier for posterity.

WILLIAM BLASIUS,
"Storms: Their Nature, Classification and Laws," 1875

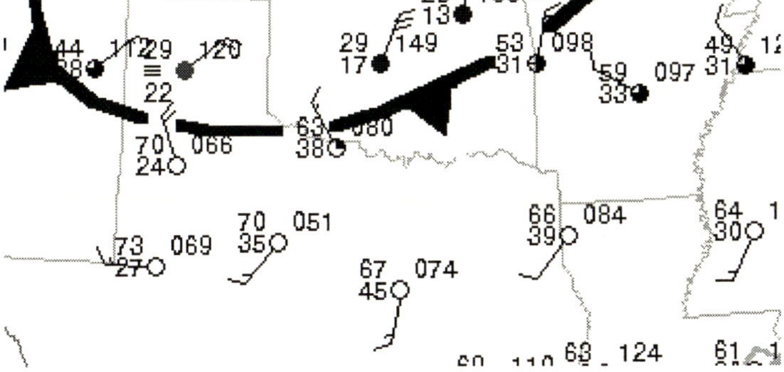

Figure 5-3. Cold front pushing south through the central Plains on January 19, 1985.

Figure 5-4. Warm front and a dryline (far lower right) on a day in March 2000 in Texas. Note how the temperatures are in the 70's south of the front, with 50's and 60's to the north with fog and haze. The front is analyzed on the warm side of the thermal gradient. The dryline, in a similar manner, is analyzed on the moist side of the moisture gradient. The "C" designators at the lower right of each plot indicate the height of the ceiling in hundreds of feet. Pressures are in units, tenths, and hundreds of an inch of mercury rather than in millibars.

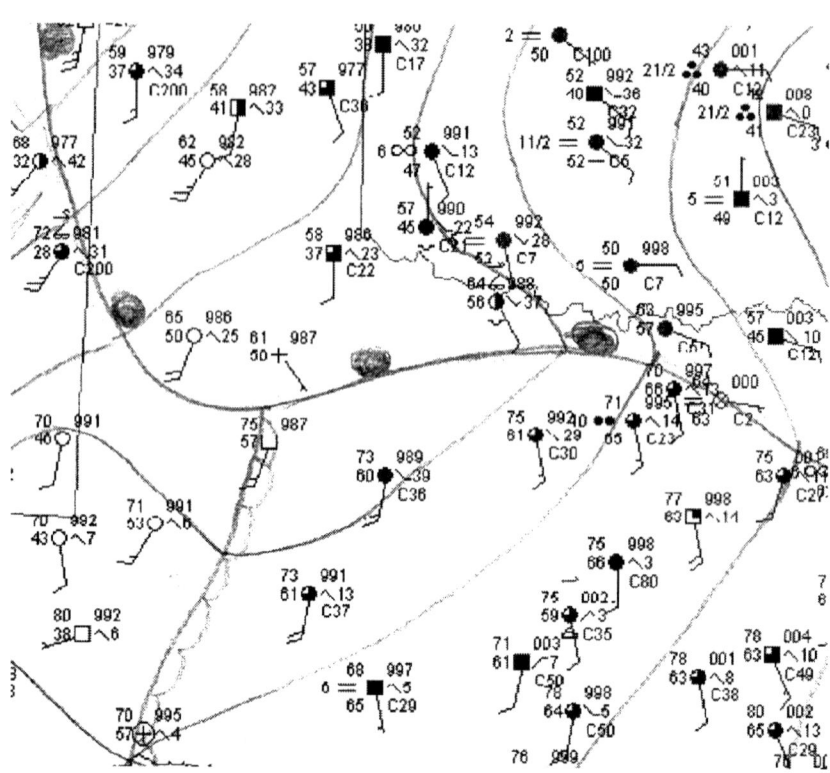

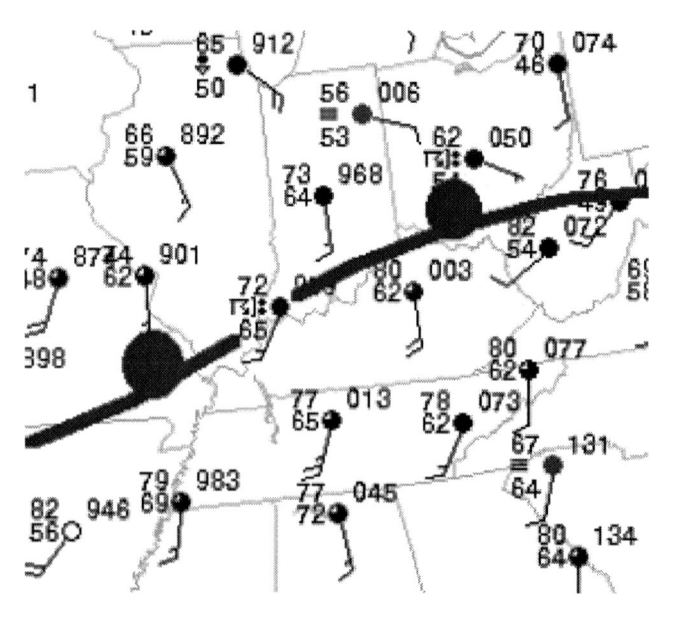

Figure 5-5. Warm front. Note that the thermal boundary has been given highest priority. The front effectively separates near-80 degree temperatures in the southern U.S. from 60s and 70s to the north. Some analysts might have chosen to place the front along the wind shift line in central Indiana, but it was the warm front shown here that became a significant factor in the evening's weather and became a focus for severe thunderstorms.

slowly after the front passes. The movement of a warm front is generally slower than that of a cold front.

5.5. QUASISTATIONARY FRONT

An quasistationary front is the term for a front in which neither air mass is replacing another. The front may dissipate or may begin moving again as a cold or warm front. For most purposes it has the characteristics of a warm front.

5.6. OCCLUDED FRONT

This is the term for a cold front which has merged with a warm front, typically due to the cold front catching up to the warm front. The occluded front begins slowly dissipating as the air masses mix around the surface low, and becomes harder to find. There are two types of occlusions.

5.6.1. Cold occlusion. When the air behind the cold front is colder than the air ahead of the warm front, the cold front will lift both air masses adjoining the warm front, leaving only the cold front at the surface. The cold front is then referred to as an occluded front (cold occlusion). This is the most common type of occlusion. The greatest depth of warm air aloft is behind the occluded front, so a pressure trough tends to occur behind the cold occlusion under the axis of deepest warm air.

5.6.2. Warm occlusion. When the air behind the cold front is warmer than the air ahead of the warm front, the air masses adjoining the cold front will ascend over the warm front, leaving only the warm front at the surface. The warm front is then referred to as an occluded front (warm occlusion). The greatest depth of warm air is ahead of the occluded front, so a pressure trough tends to occur ahead of a warm occlusion along the axis of deepest warm air. A warm occlusion is much more rare than a cold occlusion. One example where it occurs is along the Washington/British Columbia coast when an polar air mass has spilled into the Pacific and is being driven north along a warm front. When a weather system brings a cold front into contact with this warm front, the air behind the cold front is usually somewhat mild, and it ascends the cold air along the coast as a warm occlusion.

From a veteran storm chaser:
I've driven through several drylines, and in most instances there is a sudden, recognizable change in air masses at the leading edge of the boundary. The most dramatic encounter I recall was while driving east from Lubbock, Texas when suddenly it seemed I hit a wall of water vapor. The parched air mass inside the car suddenly filled up with saturated air and the windows fogged up. Looking north and south along the dryline I was able to view the cross-section of two different air masses. A sharp clear sky was west of the dryline and distant cumulus were crisp and white. In contrast, the air mass to the east was hazy and cloudy, and I could see only a few rows of yellow, fuzzy cumulus fading into obscurity. Later that day, upon my return westward, I suddenly plunged through a wall of dust denoting the leading edge of the dryline. The muggy and oppressive air suddenly evacuated the vehicle and it felt like I had driven into a blast furnace. My lips became chapped and my teeth gritty.

TIM MARSHALL,
"Dryline Storms," 1988

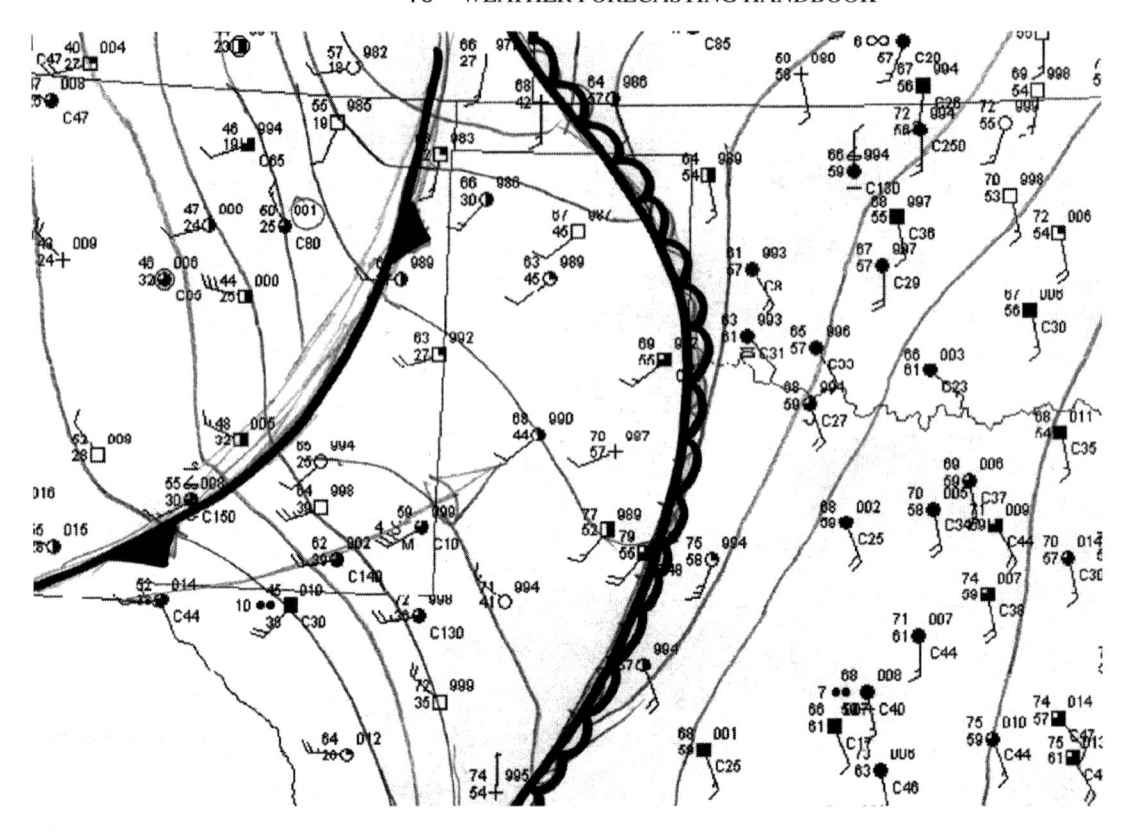

Figure 5-6. Dryline in west Texas. The dryline marks the westernmost extent of moisture. Note a cold front further west in New Mexico. This cold front was difficult to locate due to significant station elevation variations creating a spectrum of temperature readings, however an analysis of surface potential temperature was used to narrow down its location.

5.7. DRYLINE

Drylines demarcate a strong moisture gradient between warm tropical air and warm continental air. They are usually found in the Great Plains during the springtime, and may also be located in India and Australia. The dryline is significant because it is frequently associated with violent thunderstorm activity in the springtime.

5.7.1. Structure. The dryline is best visualized by looking at the moist air mass as a whole and thinking of the dryline as a boundary on the western edge. The moist air mass signifies the northward migration of tropical moisture. It intrudes into the dry continental air in the form of a shallow layer hugging the earth (anywhere from 1000-5000 ft deep or more). Therefore the dry air mass to the west of the dryline is pretty much identical to the air mass above. The dry air mass that exists above the moist sector is often referred to as the EML, or elevated mixed layer. Much of it originates from New Mexico and northern Mexico.

5.7.2. Location. The dryline is located on the moist side of a moisture gradient. The moisture gradient itself is best found using dewpoint contours, or better yet, mixing ratio contours. The dryline is not a front, since it separates air masses with different moisture values rather than different densities. For example, temperatures in the dry sector can drop to 50 deg F at night and rise to 100 deg F during the day, while the moist sector remains in the 70s and 80s. Therefore the dryline must not be located using temperature as criteria. A trough may be associated with the dryline; a cold front should not be drawn in such a trough unless it is clear that a dryline does not exist there.

5.7.3. Significance. The dryline represents the westernmost extent of significant low-level moisture. Since upper-level dynamics move from west to east, having little effect in the dry sector, the dryline is the first location where thunderstorms develop. Also since lapse rates are steeper in the dry sector, momentum from upper-level winds tends to be transported downward, shifting winds throughout the dry sector, sometimes becoming gusty and raising dust, and adding to convergence of the wind field along the dryline. This convergence is often significant in producing upward vertical motion and severe weather along the dryline. Other small-scale circulations have been identified within the dryline that relate to moisture differences and contribute to convergence along it.

5.7.4. Movement. The dryline tends to move westward at night due to advection, as the wind flow brings higher amounts of moisture to its western fringes. In these areas, the dewpoint will increase sharply at night and it will seem as if the dryline passed and moved westward. During the day, the dryline moves through mixing. As daytime heating progresses, convection tends to mix the moist air mass. Along the western fringes, much of it is dispersed into the dry layer aloft. A station in this area would note its dewpoint has fallen, and would infer that the dryline has passed and moved eastward. This daytime movement is often poorly related to the wind field and can only be predicted by examining the depth of the moist air mass and the amount of heating and mixing expected. This cyclic motion each day is sometimes referred to as "sloshing". The dryline is sometimes overtaken by a frontal system, especially late in the spring, and both the dry and moist sector are replaced with a polar air mass.

The sea breeze in Chile
In the summer of the southern hemisphere the sea breeze at Valparaiso is more powerfully developed than at any other place to which my services afloat have led me. Here regularly in the afternoon, at this season, the sea breeze blows furiously. Pebbles are torn up from the walks and whirled about the streets; people seek shelter; the Almendral is deserted, business interrupted, and all communication from the shipping to the shore is cut off. Suddenly the winds and the sea, as if they had again heard the voice of rebuke, are hushed and there is a calm. The lull that follows is delightful. The sky is without a cloud; the atmosphere is transparency itself; the Andes seem to draw near; the climate, always mild and soft, becomes now doubly sweet by the contrast.

M.F. MAURY
"The Physical Geography of the Sea," 1874

5.8. SEA/LAND BREEZE FRONTS

These fronts are the result of unequal heating between the land and large bodies of water. They are most prominent in benign weather patterns, such as during the late summer. When strong weather systems are in the area, they can be displaced by the prevailing flow or suppressed altogether.

5.8.1. Sea breeze. During the morning hours, the land heats the air above it at a much faster rate than that over the ocean, which causes it to rise. Air over the ocean then flows inland to compensate for the rising air. After a few hours, a circulation is established where air flows from sea to land at the surface and from land to sea aloft.

5.8.2. Land breeze. At nighttime, rapid cooling of the air over the land reverses the circulation, allowing air to flow from land to sea at the surface and from sea to land aloft. This circulation is much weaker than that of the sea breeze.

5.8.3. Temporal characteristics. The sea breeze usually starts in the late morning hours, peaking in the mid-afternoon and ending before midnight. The corresponding "upper land breeze" lifespan is nearly the same as the sea breeze, and is usually at an altitude of 3000 to 9000 ft. At night, the land breeze is most established just before dawn, with the corresponding "upper sea breeze" occurring in mid-morning at an altitude of about 5000 to 7000 ft.

5.9. JETS

A jet is the name given to almost any narrow band of strong winds. The most widely-known type of jet is the polar front jet, also known as the "jet stream". There are other types of jets, too, which we'll discuss here.

5.9.1. Polar front jet (PFJ). The polar front jet is created by low-level temperature contrasts in the middle latitudes. These contrasts produce different thicknesses in the atmosphere above them, intensifying the pressure gradients aloft. Winds aloft in the Northern Hemisphere flow from south to north over the contrast to fill the pressure "void", and as they accelerate they are deflected towards the east by the Coriolis force, forming a "river" of air that moves from west to east around the globe.

FRONTS AND JETS • 81

* *Characteristics.* The PFJ flows in a series of segments around the hemisphere. A hot air balloon launched into the jet stream would flow from North America to Europe, then across Asia and back over North America. The mean location of the jet stream migrates north during the summertime, when the extent of polar air masses recedes north, taking thermal contrasts with them. In North America, it is usually found over southern Canada during the summer. But in wintertime, its mean position shifts southward, placing it somewhere in the United States on any given day.

* *Significance.* The PFJ represents a major source of energy for weather systems. For the jet stream to form in the first place, a clash between polar and tropical air has to occur. But once this happens and the jet stream strengthens, the jet can give some of its dynamics back to the surface systems, helping them develop further (this will be covered in detail shortly). And as the surface systems intensify, even more energy is transferred back to the jet by thermal contrasts! It's a chain cycle that helps produce some of the large wintertime storm systems occasionally found across the United States. The jet can also give some of its dynamics to severe thunderstorms on the plains, or even to high pressure regions!

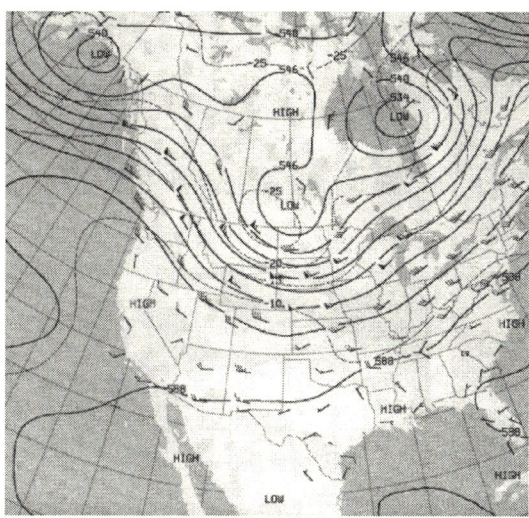

Figure 5-7a. Textbook, single jet stream pattern. The polar front jet flows from British Columbia to Wyoming, then up to Michigan and Quebec. Over the southern United States, winds are light. It is relatively easy to forecast weather when the upper-level charts are this simple. (9 September 1992)

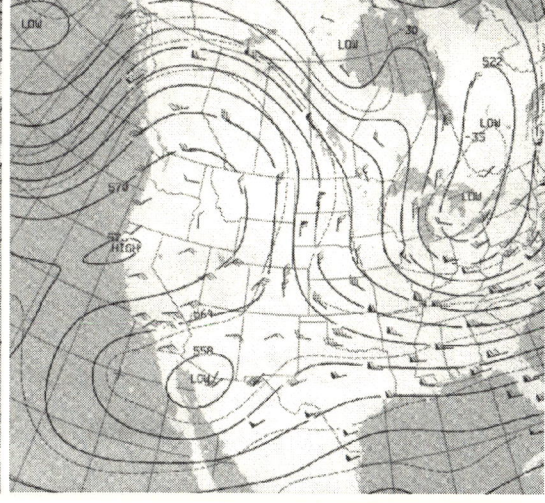

Figure 5-7b. A split flow pattern, which is more often the rule than the exception. This map shows the main polar front jet flowing from northern British Columbia to Minnesota to Virginia. The southern branch of the polar front jet flows from northern Mexico to Texas and into Alabama. (2 April 1992)

5.9.2. Subtropical jet (STJ).
The STJ is caused by slight temperature contrasts along the boundary of the Hadley and Ferrel cells in subtropical latitudes. As parts of the STJ bulge north, the air parcel flow through the jet axis at a higher velocity due to increased angular momentum (AM = mvr, mass is constant and radius of the Earth decreases, so velocity must increase).

* *Characteristics.* Usually strongest at 200 mb, the normal latitude of the STJ is about 28 deg, but can vary between 25 and 35 deg. It is stronger in the winter than in the summer.

* *Significance.* The STJ is often a key player in transporting warm, moisture-laden air northward, enhancing mid-latitude weather systems.

* *Features.* Cirrus is often found on the warm side of the STJ axis, sometimes forming transverse bands.

5.9.3. Low-level jet (LLJ).
The low-level jet is a band of unusually strong southerly winds (south-to-north) in the lower levels of the atmosphere, usually seen ahead of developing cyclones on the Great Plains. They are strengthened by the indirect transverse circulation ahead of a polar front jet (discussed later under "jet streak dynamics").

* *Characteristics.* Low-level jets are strongest at 850 mb or 925 mb during the overnight hours. The typical location is from east Texas northward to Kansas or Missouri.

* *Significance.* Low-level jets are important in advecting large volumes of warm, moist air northward from the Gulf of Mexico into the central United States. This process is a major factor in the release of severe thunderstorms later during the afternoon and evening hours.

* *Features.* The LLJ may be associated with stratus clouds along and south of the jet's nose. In the deeper moisture, stratocumulus layers may be observed.

REVIEW QUESTIONS

1. Which is most likely to cause turbulence near the ground: a cold cPk air mass moving south over warmer terrain, or a moist, hot mPw air mass moving north over cold terrain?

The low level jet

In Dallas, a late night in April can be an eerie experience. As powerful spring systems approach Texas and local forecasters talk about tomorrow's severe weather, the low-level jet quietly gathers strength. While stargazing and enjoying the calm, quiet night, you notice a few fragments of stratus cloud moving in rapidly from the south. These clouds gradually become more dense, forming a thick overcast that races northward at dizzying speeds. The night is as still and quiet as ever, but as you look at the overcast cloud moving swiftly overhead, you can almost hear a distant roar. Is it the low-level jet or is it the imagination?

TIM VASQUEZ

FRONTS AND JETS • 83

2. What is the primary measurable characteristic of a front?

3. Is surface convergence or divergence more likely to strengthen a frontal boundary?

4. You are hanging out at the weather station in Omaha, Nebraska. After the balloon launch, you and the technician note that there is a frontal inversion 2000 ft above the surface. Is the front north of you (approaching) or south of you (already passed)?

5. When analyzing a thickness chart, you notice that the thickness lines cross the surface cold front at a sharp angle (nearly perpendicular). What does this reveal about the cold front?

6. What is the primary measurable characteristic of a dryline?

7. What period of day (morning, day, evening, night) is mixing a dominant process in dryline behavior?

8. With summer coming your buddy Earl has a plan to build a bar & grill just inland from the beach. His clever scheme is to barbecue a wide variety of meats in the late afternoon and let the aromas waft with the wind onto the beach, attracting the dinnertime crowd. Is Earl's plan ingenius or a dud?

9. You have a plan to launch a round-the-world hot air balloon attempt, and you need a strong upper-level wind. Knowing the jet stream is common in Canada in one season and in the United States another, is it better to launch in the summer or winter?

10. What type of weather phenomena is closely linked to the low-level jet?

Altocumulus castellanus in Norman, Oklahoma. *(Tim Vasquez)*

6 MOTION

*Great whirls have little whirls,
That feed on their velocity,
And little whirls have smaller whirls
And so on to viscosity.*

L.F. RICHARDSON, envisioning large-scale winds creating a cascade of energy transfers through smaller and smaller scales down to molecules, where it finally changes to heat.
Weather Prediction by Numerical Process, 1922

The atmosphere across the globe, being a fluid, is not at all unlike the ocean, comprised of waves and eddies. A parcel of air travelling around the world can never travel in a straight line! These discontinuities in the wind fields are caused by imbalances in the pressure fields, which in turn are caused by uneven heating of the Earth.

Forecasters are quite interested in the nature of these waves because, depending on the scale and size, they dictate anything from weather during the next hour to weather during the upcoming week. To understand them, it's necessary to start at the largest possible scales and work our way down.

6.1. LONG WAVES

High in the atmosphere, the winds and heights are organized into very large waves measuring thousands of miles in size. These are called long waves, or "Rossby waves". They are usually masked by smaller-scale waves embedded within them. On a typical day there are four or five long waves encircling the globe in the temperate latitudes of both the northern and the southern hemisphere.

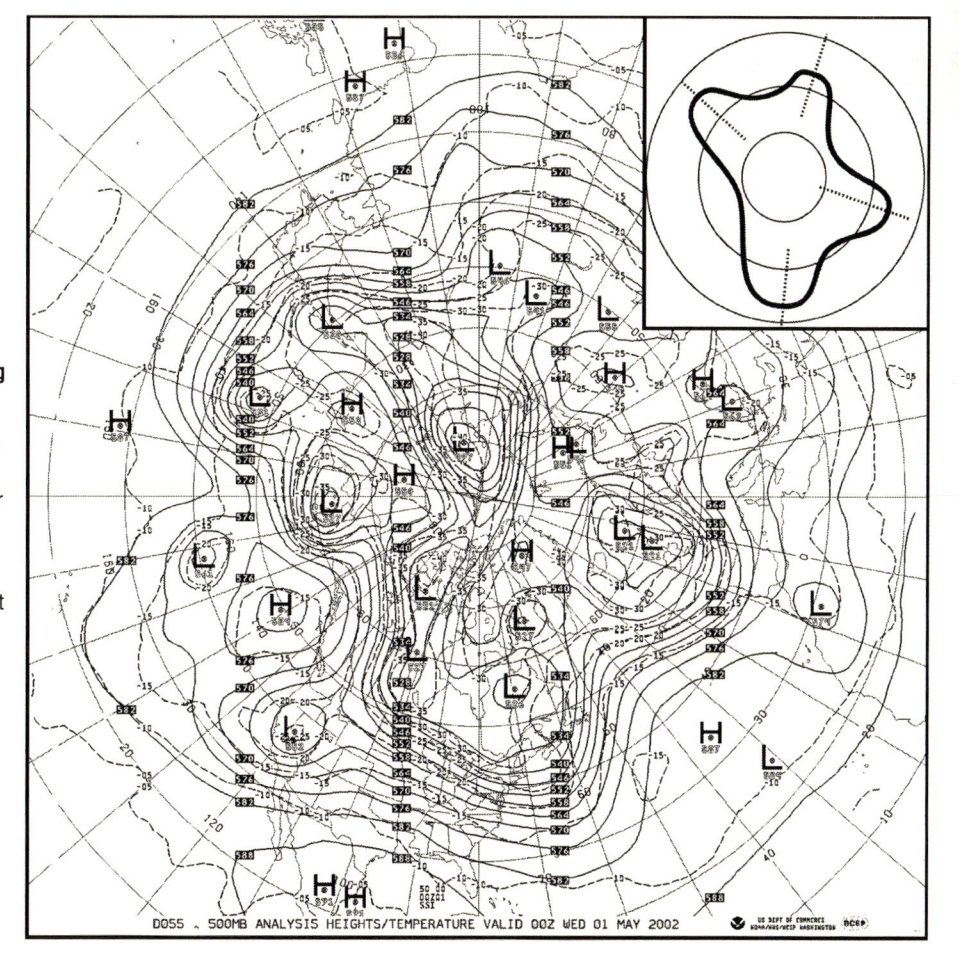

Figure 6-1. Long waves can be picked out on hemispheric 500 mb charts. This chart shows four long waves comprised of smaller-scale waves. The inset shows the long waves, with the trough axes indicated by dashed lines.

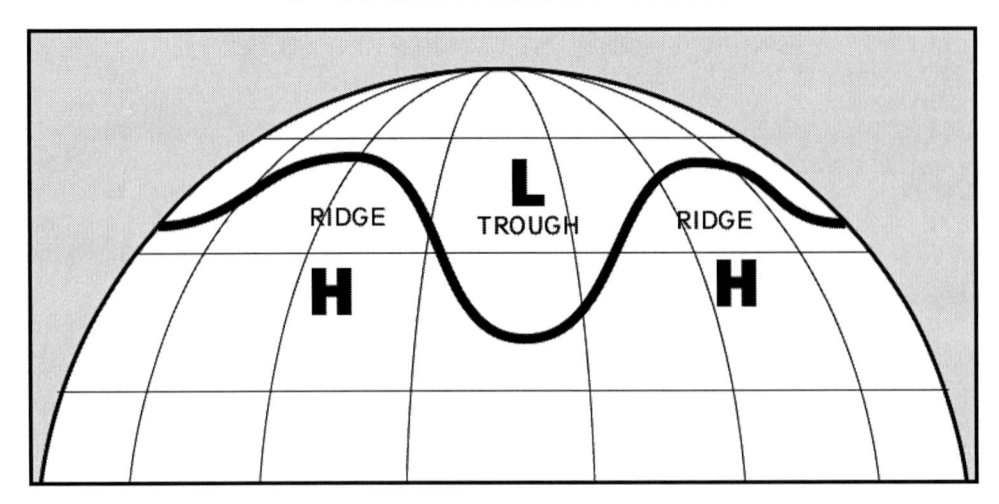

Figure 6-2. Troughs and ridges in the upper troposphere. Since the polar regions contain low pressure (low heights), any southward expansion of contours and flow from the pole defines a trough. The opposite is true for a ridge. If you're thinking that upper-level troughs are associated with cold polar air masses, you are indeed correct.

<u>6.1.1. Scale</u>. Long waves take up a scale the size of a continent. They are about 50 to 120 degrees of longitude in wavelength, have an amplitude of 1500 to 1800 miles, and are best seen at 300 and 200 millibars (above 30,000 feet). Most of the irregularities seen in the lower layers have dampened out at these heights, and you'll see broad, large-scale ridges and troughs.

<u>6.1.2. Characteristics</u>. There are usually 4 or 5 long waves around the hemisphere, but there can be as many as 7 or as few as 3. Their amplitude is 1500 to 1800 miles, and they have a wavelength of 50 to 120 degrees of longitude. They move eastward at up to 15 knots or may be stationary or retrogress at up to 3 knots.

<u>6.1.3. Impact</u>. Long wave ridges are generally associated with warm weather, with the area downstream typically experiences fair weather.. Long wave troughs are associated with cold temperatures, and the area downstream typically gets unsettled or stormy weather, especially if the long wave trough is deep enough to dip into subtropical moisture (which often occurs in the southern U.S. in the wintertime).

<u>6.1.4. Number</u>. When long waves are few, they will have larger wavelengths and will move slower. As a general rule, if there are only 3 long waves, the pattern around the hemisphere will retrogress. Retrogression is rare but it does occur from time to time. At 4, the pattern will remain stationary or move slowly, and at 5 or 6, the long waves will progress from west to east.

6.1.5. Patterns. There are various patterns that are associated with long waves:

* *Seasonal patterns.* During the winter, a long wave trough pattern tends to cover a large part of continental landmasses while ridges prevail over the open ocean. This is because the continental area has a lower mean temperature, with lower vertical thicknesses that translate to lower heights aloft. The opposite effect occurs during the summer season.

* *Polar vortex.* A deep upper-level low is usually present over northeastern Canada during the winter. It is called the polar vortex. It is a reflection of low upper-level heights over the most frigid parts of the continent. It may occasionally be displaced southward across southeastern Canada, the Great Lakes, and New England. When this happens, rapid-moving short wave troughs moving through the long wave pattern can produce intense storms with strong surface winds and extensive precipitation across the northeast United States.

6.1.6. Long wave ridges. A long wave ridge will build when warm air advection or a major short wave ridge (to be covered shortly) moves into the long wave ridge. A jet streak moving into the back side of a long wave ridge will cause it to build, and if there is a deep trough further downstream, the low may close off.

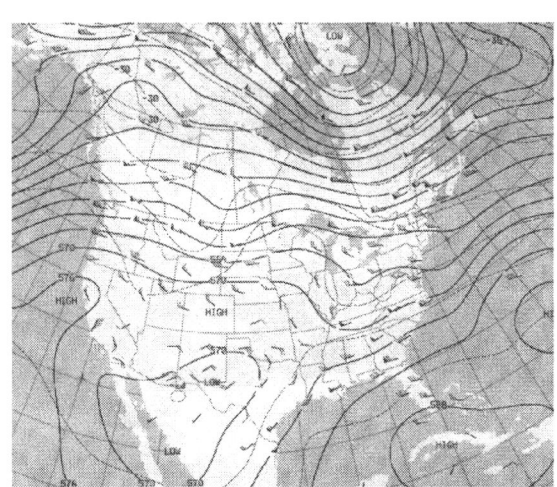

Figure 6-3a. High-zonal pattern. Note how the flow is predominantly west-to-east, with only minor variations. This pattern is generally unstable in nature and the upper-level flow soon tends to break up into a series of waves. A series of weak frontal systems was moving across the Great Lakes region, where the jet stream was stronger. (27 January 1992, 500 mb)

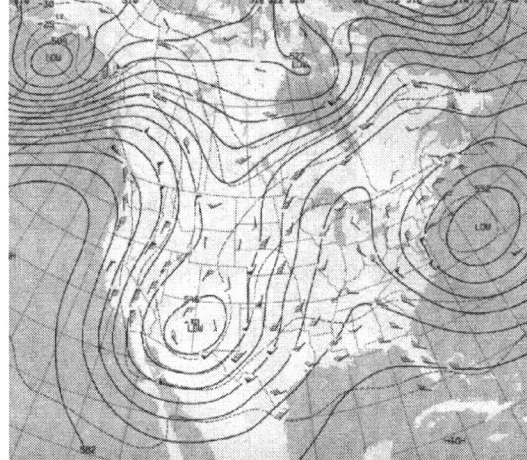

Figure 6-3b. Low-zonal pattern. This generally represents a breakup and cellular fragmentation of the upper-level flow. Note how the flow is almost direct northerly over the Pacific coast, turning sharply to a southwest flow over the Great Plains. Winter precipitation fell over the central United States during the next couple of days, while heavy rains and thunderstorms swamped east Texas and Louisiana. (13 December 1992, 500 mb)

Long waves
* Long wave troughs and ridges may be difficult to determine from a single 500 mb chart.
* Eighty percent of the time long wave patterns do not move.
* Short wave troughs deepen at the base of the long wave position.
* Short wave troughs can flatten mean ridges resulting in a zonal chart.
* Surface cyclogenesis takes place east of stationary long wave trough axes. The surface low tracks northeast at this time.
* Short wave troughs tend to be diffluent upstream and confluent downstream of the long wave position.

S. BUSINGER,
"The Long Wave Concept", 1987

A jet streak moving out of the ridge will cause it to weaken. If a jet streak is moving around a sharp ridge and a northeast-southwest long-wave trough is downstream, the trough will fill and orient from north-to-south. If a jet streak is moving around a sharp ridge and a north-south trough is downstream with a blocking ridge further downstream, the trough will fill.

6.1.7. Long wave troughs. Long wave troughs will deepen when cold air advection or a short wave trough moves into the long wave. A jet streak moving into the back side of a long wave trough will deepen it, while a jet streak moving out of it will cause it to fill.

6.2. SHORT WAVES

6.2.1. Origin. Short waves are small-scale waves associated with temperature advection. They show up best at 700 mb, but many of the larger ones may extend up to 500 mb.

6.2.2. Scale. Short wave troughs and ridges are about the size of several states or less, and are imbedded in the long-wave pattern. They have a wavelength of 1 to 40 degrees of longitude (those over 15 degrees are considered "major short waves"), and have an amplitude averaging 100 to 1000 miles. They are small, and can create intense vertical motion, often indicated by PVA or NVA (discussed later).

6.2.3. Location. Short wave troughs are best found by locating a cyclonic wind shift in the prevailing flow and/or a thermal trough upstream of the short wave. Short wave ridges can be found by locating an anticyclonic wind shift and/or a thermal ridge upstream of the short wave. Positive vorticity at 500 mb can help locate short waves; the highest units of vorticity tend to reflect the location of the short wave.

6.2.4. Mechanics. Short waves are not really an object, but a process. Low-level warm air advection ahead of the short wave maintains the slightly higher heights ahead of it (remember warmer air has more thickness, resulting in higher upper-level heights), whereas low-level cold-air advection maintains the short-wave trough itself.

6.3. DIVERGENCE/CONVERGENCE

6.3.1. Divergence. The divergence of the wind field is a measure of the rate of net removal of mass out of a volume of air above a

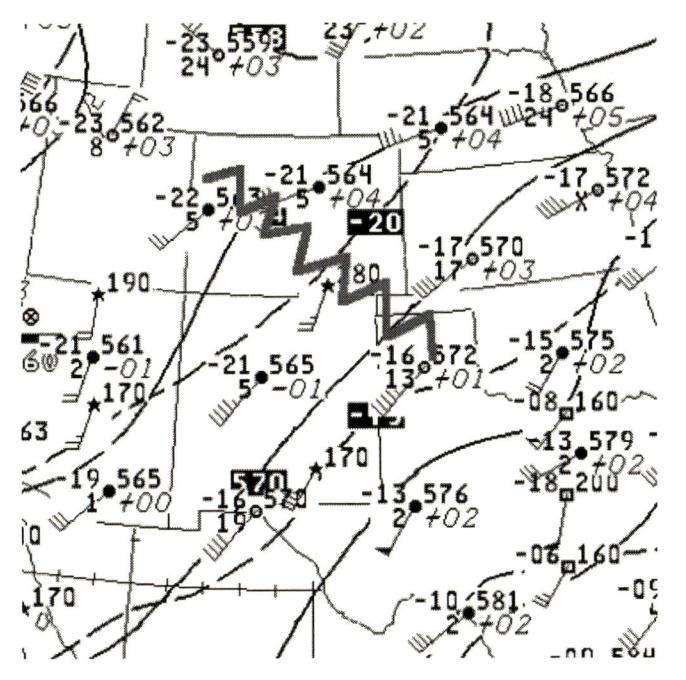

Figure 6-4. A "short wave" does not always mean a short wave trough. Here we see a short wave ridge. It is drawn here with the jagged line and is placed where anticyclonic turning of the winds is present.

given point. It results in a decrease of atmospheric mass above a given point, and thus, pressure falls. This is because the equations for force and pressure show that removing mass decreases force, and decreasing force decreases pressure. Divergence can be evaluated by examining difluence and speed divergence.

* *Directional divergence (difluence)*. Difluence is one component of divergence. It is the directional spreading of wind flow. So if speed of the wind is constant everywhere but the flow shows difluence, divergence is occurring.

* *Speed divergence*. Speed divergence is another component of divergence. It is easiest to illustrate by picturing a highway with three lanes of traffic going 50 mph, and ahead of those cars is three lanes of traffic going 70 mph. In a sense, "mass" is being removed from between these vehicles as they spread apart. In the atmosphere, this is known as speed divergence.

* *Final evaluation*. When looking at charts, both difluence/ confluence and speed divergence/convergence must be looked at separately to determine if divergence is occurring. For example, on the front side of a jet streak it is common for difluence to be

---

Before chaos was recognized
I remember a talk that John von Neumann [great mathematician] gave at Princeton around 1950, describing the glorious future which he then saw for his computers. Meteorology was the big thing on his horizon. He said, as soon as we have good computers, we shall be able to divide the phenomena of meteorology cleanly into two categories, the stable and the unstable. The unstable phenomena are those which are upset by small disturbances, the stable phenomena are those which are resilient to small disturbances. He said, as soon as we have some large computers working, the problems of meteorology will be solved. All process that are stable we shall predict. All processes that are unstable we shall control. He imagined that we needed only to identify the points in space and time at which unstable processes originated, and then a few airplanes carrying smoke generators could fly to those points and introduce the appropriate small disturbances to make the unstable processes flip into the desired directions. The dream was based on a fundamental misunderstanding of fluid motions [which] fall into a mode of behavior known as chaotic. A chaotic motion is generally neither predictable nor controllable. It is unpredictable because a small disturbance will produce exponentially growing perturbation of motion. It is uncontrollable because small disturbances lead only to other chaotic motions and not to any stable and predictable alternative.

FREEMAN DYSON
Infinite In All Directions, 1988

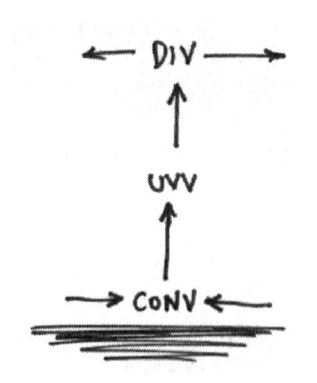

Figure 6-5a. The chimney effect. Divergence aloft and/or convergence at the surface will produce upward vertical motion (UVV's, or "upward vertical velocities").

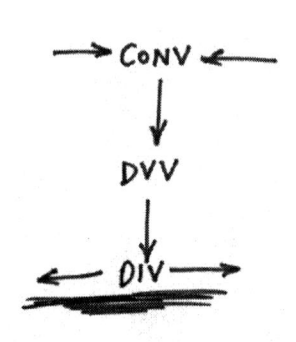

Figure 6-5b. The damper effect. Convergence aloft and/or divergence at the surface will produce downward vertical motion (DVV's, or "downward vertical velocities").

cancelled out by speed convergence. The best way to tell whether divergence is occurring is through careful examination of both elements, or to use a computer program (such as Digital Atmosphere) that mathematically analyzes the wind field.

<u>6.3.2. Convergence</u>. The convergence of the wind field is a measure of the rate of net addition of mass into a volume of air above a given point. It results in an increase of atmospheric mass above a given point, and thus, pressure rises. This is because the equations for force and pressure show that adding mass increases force, and increasing force increases pressure. Convergence can be evaluated by examining confluence and speed convergence.

* *Directional convergence (confluence)*. Confluence is one component of convergence. It is the directional merging of wind flow. So if speed of the wind is constant everywhere but the flow shows confluence, convergence is occurring.

* *Speed convergence*. Speed convergence is another component of convergence. It is easiest to illustrate by picturing a highway with three lanes of traffic going 60 mph, and ahead of those cars is three lanes of traffic going 40 mph. In a sense, "mass" is being added between these vehicles as they close in on each other. In the atmosphere, this is known as speed convergence.

* *Final evaluation*. When looking at charts, both confluence/difluence and speed convergence/divergence must be looked at separately to determine if convergence is occurring. For example, on the rear side of a jet streak it is common for confluence to be cancelled out by speed divergence. The best way to tell whether convergence is occurring is through careful examination of both elements, or to use a computer program (such as Digital Atmosphere) that mathematically analyzes the wind field.

6.4. VERTICAL MOTION

Many introductory weather books tend to gloss over vertical motion. However, this type of motion is critical because it directly affects what the weather is going to be like. Meteorologists are always trying to evaluate and predict vertical motion every day. In synoptic scale motion, winds aloft are on the order of 50 meters per second, while the vertical motion that produces clouds and rain is on the order of 1 centimeter per second. This is impossible to measure directly, so forecasters rely on other indicators to analyze and predict vertical motion.

6.4.1. Rising motion. If upper-level divergence occurs, the troposphere will attempt to compensate by initiating rising motion (to fill the "void"). Increased convergence in the lower troposphere will usually result. This process is sometimes referred to as the "chimney effect".

* *Weather*. When air is forced to rise vertically from the surface, it rises into regions where pressure is lower (remember that pressure decreases with height). As a result, the volume of the air parcel expands, allowing its temperature to decrease. But eventually, the temperature will decrease to a point at which water vapor will be squeezed out of it. This produces visible cloud droplets, causing clouds. Continued rising motion produces more clouds, and the droplets can agglomerate into larger precipitation droplets, which are heavy enough to fall out of the cloud.

* *Pressure falls*. If the divergence aloft is stronger than the convergence in the lower levels, surface pressure falls will occur (mass is being removed from the column by the divergence).

* *Stratospheric warm sinks*. If divergence aloft is strong enough, slight sinking motion may occur in the lower stratosphere, resulting in adiabatic warming. The result is warmer temperatures that are sometimes found above strong divergence areas (at 200 mb). This is called a "warm sink".

Misuse of difluence
Difluence is simply the spread of streamlines downstream . . . The difluence is given only by part of the first term [for divergence], therefore, the difluence of the flow cannot be equivalent to the divergence. It is a common mistake, however, to equate difluence with divergence (and then go on to infer vertical motion, another mistake).

CHUCK DOSWELL,
"Pet Peeves of Chuck Doswell," 2000

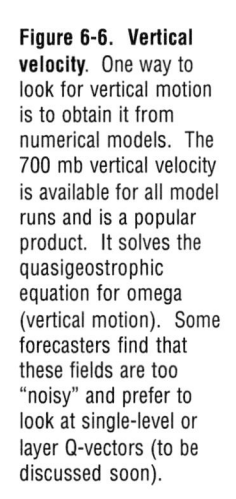

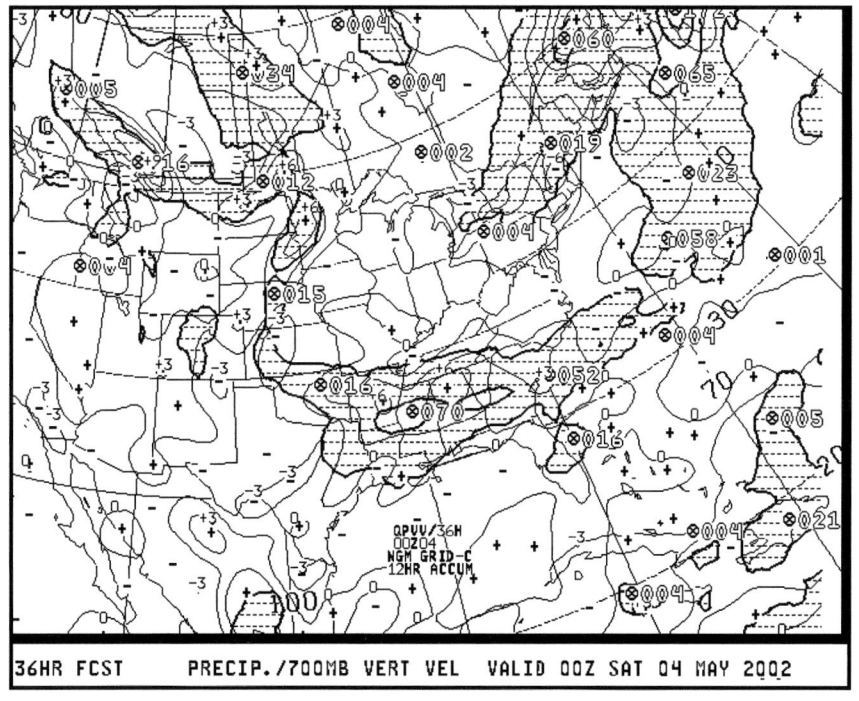

Figure 6-6. Vertical velocity. One way to look for vertical motion is to obtain it from numerical models. The 700 mb vertical velocity is available for all model runs and is a popular product. It solves the quasigeostrophic equation for omega (vertical motion). Some forecasters find that these fields are too "noisy" and prefer to look at single-level or layer Q-vectors (to be discussed soon).

Why is forecasting so difficult?
Consider a rotating spherical envelope of a mixture of gases, occasionally murky and always somewhat viscous. Place it around an astronomical object nearly 8000 miles in diameter. Tilt the whole system back and forth with respect to its source of heat and light. Freeze it at the poles of its axis of rotation and intensely heat it in the middle. Cover most of the surface of the sphere with a liquid that continually feeds moisture into the atmosphere. Subject the whole to tidal forces induced by the sun and a captive satellite. Then try to predict the conditions of one small portion of that atmosphere for a period of one to several days in advance.

Author unknown

6.4.2. Sinking motion. If upper-level convergence occurs, the troposphere will attempt to compensate by initiating sinking motion, or subsidence (to drain the "excess"). Increased divergence in the lower troposphere will usually result. This process is sometimes referred to as the "damper effect".

* *Weather*. The sinking motion produces adiabatic warming, and often clearing weather and fair skies.

* *Pressure rises*. If the convergence aloft is stronger than the divergence in the lower levels, surface pressure rises will occur (mass is being added to the column by the convergence).

* *Stratospheric cold domes*. If convergence aloft is strong enough, slight rising motion may occur in the lower stratosphere, resulting in adiabatic cooling. The result is cooler temperatures that are sometimes found above strong convergence areas (at 200 mb). This is called a "cold dome".

6.4.3. Level of non-divergence. Dines Compensation Principle states that there must be at least one level of nondivergence in the troposphere. This level is called the level of non-divergence (LND). It is usually around 550 mb, but can be highly variable depending on atmospheric stability. If convergence occurs above the LND, divergence occurs below it, and vice versa.

6.4.4. Frictional effects. In the boundary layer, frictional effects can produce vertical motion.

* *Frictional convergence produces rising motion*. This is a form of boundary layer convergence. An example is winds blowing across Lake Michigan and into Michigan itself; friction is less over the lake than on land, so the air slows over Michigan and convergence between the lake and the land wind results.

* *Frictional divergence produces sinking motion*. This is a form of boundary layer divergence. It can include wind coming out of a mountainous area and onto flat terrain; friction decreases over the flat area and the wind speeds up. The area between the slow mountain wind and the fast plains wind is an area of divergence.

6.4.5. Forcing from terrain. Mountains or even sloped land can produce vertical motion.

* *Upslope flow produces rising motion*. An example of upslope flow is a day in Kansas where a humid east wind is blowing. That night, stratus and fog begins developing.

MOTION • 93

* *Downslope flow produces sinking motion.* An example is a westerly wind blowing across the Rocky Mountains into the foothills. This air sinks and clouds dissipate.

6.5. JET STREAK DYNAMICS

6.5.1. Jet streak entrance. Air flowing into a jet streak is moving slowly. However, the height gradient begins to tighten, increasing the air's tendency to flow toward lower heights. So it begins turning slightly in that direction (to the left). But since the air is moving slowly, the Coriolis effect is still weak. At this point, the height gradient force is greater than the Coriolis effect, so this air is called "subgeostrophic". As the air continues moving into the lower heights, it begins, re-establishing the geostrophic balance.

Figure 6-7. Jet streak quadrants using a real example for the Pacific Coast on the evening of May 5, 2002. The core of the jet is easily defined by the bullseye pattern of shaded isotachs. This example shows the different quadrants with their standard names. The abbreviations are RRQ: right rear quadrant; LRQ: left rear quadrant; LFQ: left front quadrant; RFQ: right front quadrant. This is a jet stream in cyclonic curvature, and as we shall soon see, the left quadrants have definite vertical motion while the right quadrants have indeterminate vertical motion. Therefore any ships plying the Pacific away from the Canadian coast shouldn't have to worry about any surprise weather.

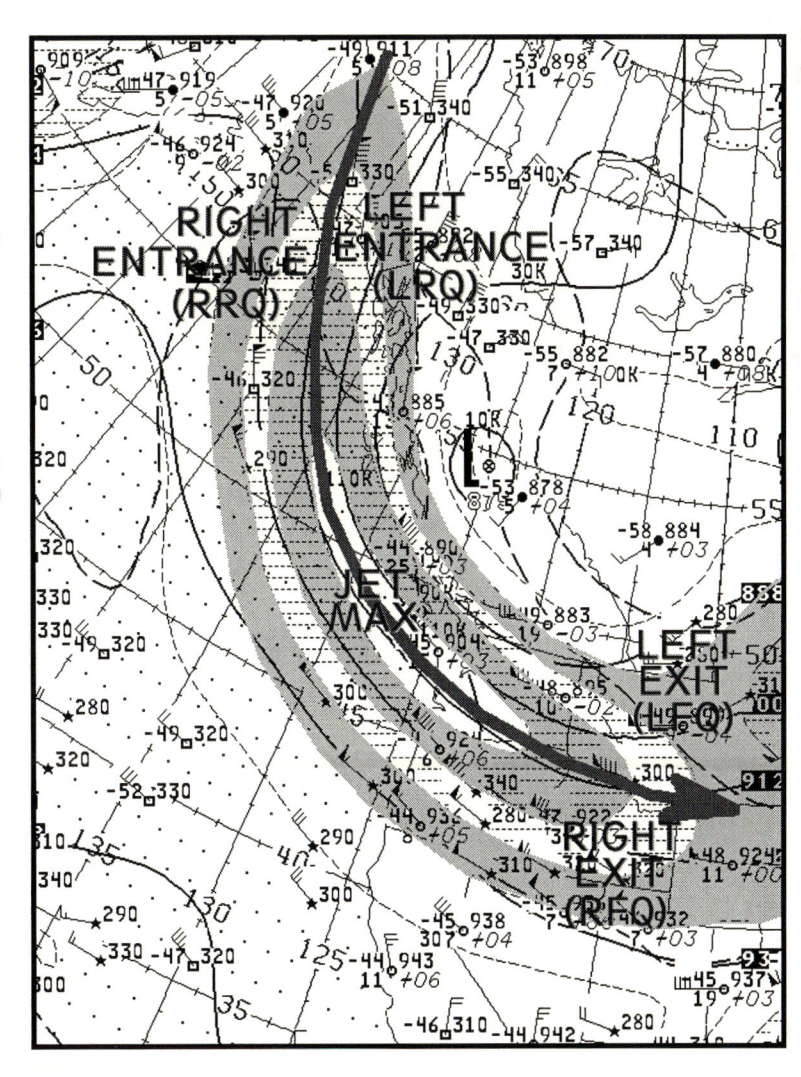

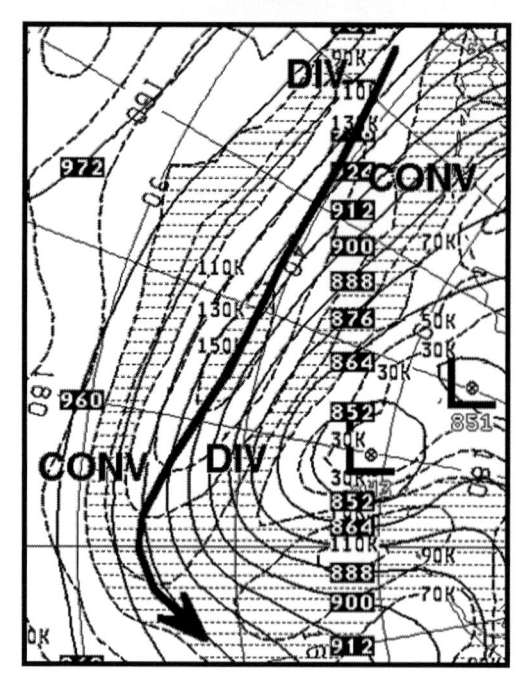

Figure 6-8. Jet streak ageostrophic flow. The core of the jet is defined by the bullseye pattern of isotachs. Note how in the entrance (top) the air seems to cross the contours to lower heights. This is an area of subgeostrophic flow, and results in convergence on the left side of the jet (as one faces downstream). In the exit side (bottom), the air is obviously crossing contours to higher heights. This is an area of supergeostrophic flow, and results in convergence on the right side of the jet (as one faces downstream). The convergent areas are associated with fair weather while divergent areas are associated with cloudy, unsettled weather.

6.5.2. Jet streak exit. When air is flowing through a jet streak core, it is moving at a high velocity, therefore, the Coriolis effect is strong. The height gradient is also tight, so the height gradient force is strong. These two effects are in balance, so the flow is geostrophic. However, as the air exits the jet streak, the height gradient begins diminishing. The air continues moving rapidly, so the Coriolis effect stays the same. The forces are out of balance, with the Coriolis effect dominant, and this is said to be "supergeostrophic flow". When the air makes this turn, it creates a divergent area directly ahead of the jet streak (since the air is turning away from this spot in favor of those higher heights). This divergence means that mass is being removed from the atmosphere at that location, and since surface pressure equals the weight of the atmosphere and you are removing mass, the surface pressures drop.

6.5.3. Divergent/convergent quads. This ageostropic (non-geostrophic flow) at the entrance and exit of the jet streak produces some interesting and very important effects. It produces divergent and convergent quadrants around the jet streak, which cause vertical motions.

Figure 6-9. Divergent and convergent quads of a jet max, and the associated transverse circulations. This model is frequently referred to in official weather forecast discussions in context with real weather systems.

* *Rear of jet*. Consider the entrance region (rear side) of a jet streak. The flow is subgeostrophic, and air piles up on the left side of the jet entrance (i.e. convergence). By contrast the right side is a divergent area. The result is that the LEFT REAR quadrant of a jet is CONVERGENT, and the RIGHT REAR quadrant of a jet is DIVERGENT. Since upper level convergence results in downward motion, the left rear quadrant contains subsidence, and the right rear quadrant contains rising motion.

* *Front of jet.* In the jet streak exit (front side), the flow is supergeostrophic and air piles up on the right side of the axis. So the right front quadrant is convergent, and the left front quadrant is divergent. It follows that the right front quadrant contains subsidence, while the left front quadrant contains rising motion.

6.5.4. Transverse circulations.
Transverse circulations usually set up whenever the maximum wind speed within the jet streak exceed the speed of movement of the jet streak itself. This is often the case during the springtime when a fast-moving weather system passes by and the trailing stationary front falls under the departing right-rear quadrant.

* *Indirect transverse circulation.* Air subsiding on the front right side of a jet sinks to the low levels and tends to flow along the surface to the area of rising air under the left front quadrant. It then flows out of the divergent upper-level area to the convergent right front side. This circulation is called an indirect transverse circulation. The air that is being forced to rise in the left front quadrant is relatively cold (since it is on the poleward side of the jet). Kinetic energy is being converted into available potential energy.

* *Direct transverse circulation.* In the entrance (rear side) to a jet streak, the air sinks on the left side, then flows south along the surface to the right side. This is called a direct transverse circulation. The air that is being forced to rise is relatively warm (since it is on the equatorward side of the jet), so available potential energy is being converted into kinetic energy.

6.6. THERMAL ADVECTION

6.6.1. Concept.
Thermal advection in the atmosphere is related to vertical motion. It is usually evaluated at or below the gradient level.

* *Cold air advection (CAA).* The presence of CAA is associated with downward vertical motion, surface pressure rises, and height falls.

* *Warm air advection (WAA).* The presence of WAA is associated with upward vertical motion, surface pressure falls, and height rises.

Garlic roast potatoes

5-6 medium red potatoes
1/2 c. olive oil
1/2 c. melted butter
2 tbsp garlic powder
1/4 tsp red pepper powder
1/4 tsp salt
1/4 tsp coarse black pepper
2 dashes cinnamon powder
dash paprika

Preheat oven to 400° and grease baking sheet lightly. Scrub potatoes and cut, unpeeled, into 1-2 inch chunks. Toss everything in deep bowl until all potatoes are coated. Distribute coated potatoes on pan (save sauce) and bake. Every 20 minutes remove from oven, drain excess oil, brush liberally with more sauce, and return to oven. Total baking time is 60-90 minutes or until potatoes are slightly charred.

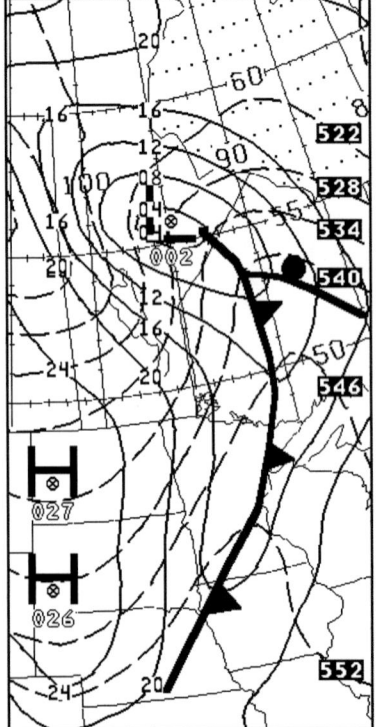

Figure 6-10. Thickness patterns, shown here as dashed lines, look surprisingly like isotherms. This isn't a coincidence: thickness is functionally like an "average isotherm" throughout the layer, in this case 1000-500 mb (surface to 18,000 ft). Therefore they shouldn't be considered intimidating. The sea-level pressure field is superimposed to help determine areas of cold and warm advection. Note how these two fields form "boxes" where advection is taking place. The smaller the box, the stronger the advection. Strong cold air advection is taking place behind the cold front, while strong warm advection is occurring north of the warm front. This example is for the morning of May 4, 2002.

6.6.2. Thickness. Thickness is the vertical distance between two constant pressure surfaces. Most often the two surfaces are 1000 mb (near the earth's surface) and 500 mb (in the middle levels of the troposphere). Charts showing thickness can be used to determine thermal advection.

* *Temperature.* The thickness of a column of air is influenced by the mean virtual temperature of the layer. The warmer the column, the higher its thickness. The colder the column, the lower its thickness. It's a lot like standing a steel beam on its end. If you heat the beam, it will expand and grow taller (thicker). If you chill it, it will shrink. Estimating the temperature of the atmosphere using thickness is similar to estimating the temperature of the steel beam by observing its thickness. The thickness of the 1000-500 mb layer in meters can be calculated by multiplying the mean virtual temperature throughout the layer (in deg K) by 20.302.

* *Moisture.* Moisture increases thickness slightly (since it contributes slightly to the virtual temperature of the air), however thickness is associated primarily with temperature.

* *Units.* Thickness is usually measured in decameter (dam) between the two pressure surfaces. A typical 1000-500 mb thickness is 548 dam (5480 m, or about 3 miles).

* *Advection.* The presence of warm air advection (WAA) is indicated where geostrophic winds (winds along isobars or isoheights) blow higher thickness values into a region occupied by lower thickness values. The presence of cold air advection (CAA) is indicated where geostrophic winds blow lower thickness values into a region occupied by higher thickness values. Neutral advection occurs if the geostrophic wind is weak (weak pressure gradient), a weak thickness gradient exists, or the geostrophic wind blows parallel to the thickness contours.

* *Advection boxes.* On a chart where sea level pressure contours are overlaid with thickness contours, advection can be determined by visually forming boxes created by the intersection of the two contour types. Areas with lots of small boxes show that strong thickness advection is taking place.

6.7. VORTICITY

Vorticity defined is a measure of spin of the air, measured in radians per second (1 revolution is 6.28 radians). For much of the 1960s, 1970s, and 1980s before computing power was available to fully diagnose vertical motion, forecasters extensively used vorticity as a key to estimating where vertical motion was occurring. First let's define the terminology, then examine the components of vorticity and see how to apply it.

6.7.1. Positive vorticity. Cyclonic spin. A spot with the greatest positive vorticity is usually marked with "X".

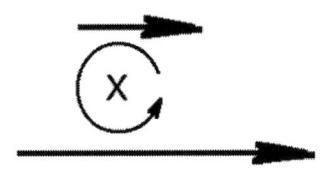

Figure 6-11a. Shear, a difference in wind speed over a given distance, causes vorticity.

6.7.2. Negative vorticity. Anticyclonic spin. A spot with the greatest negative vorticity is usually marked with an "N".

6.7.3. Positive vorticity advection. A location where the vorticity is becoming more positive (cyclonic) with time.

6.7.4. Negative vorticity advection. A location where the vorticity is becoming more negative (anticyclonic) with time.

6.7.5. Components of vorticity. There are two components that make up vorticity.

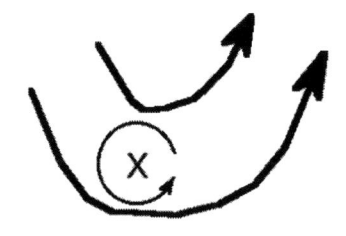

Figure 6-11b. Curvature in the airflow causes vorticity.

* *Shear.* This is vorticity resulting from lateral speed shear in the wind flow. For example, if you are driving in the right lane of the freeway at 50 mph and someone next to you in the left lane is

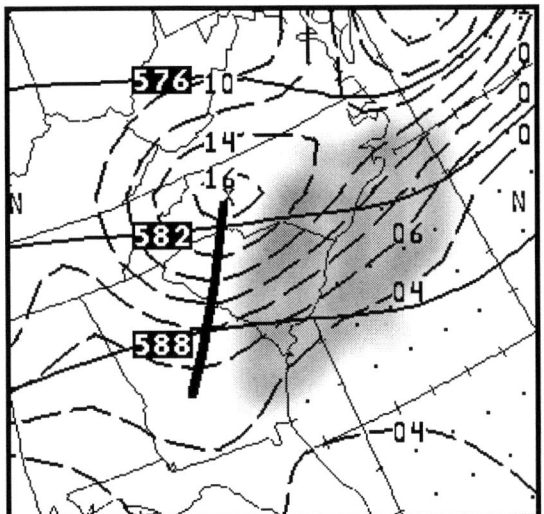

Figure 6-12a. Positive vorticity advection. As the vorticity lobe (black line) moves eastward along the 500 mb contours, higher values of vorticity are approaching the shaded area. Therefore the area is said to be in an area of "positive vorticity advection". Generally PVA can be found anyplace that vorticity lines cross the contours with higher values upstream.

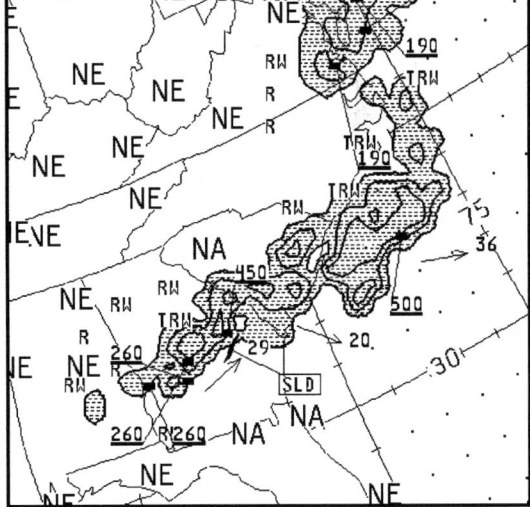

Figure 6-12b. Corresponding radar image. The upward synoptic-scale lift over South Carolina, southeastern Georgia, and southern North Carolina creates an environment favorable for development and persistence of thunderstorms.

driving 80 mph, an area of shear exists between the two cars. The same principle applies in airflow.

* *Curvature.* Vorticity resulting from the curvature of the wind flow. Regions where the air turns cyclonically generates positive (cyclonic) vorticity, while regions where the air bends anticyclonically generates negative (anticyclonic) vorticity.

6.7.6. Types of vorticity. There are also two types of vorticity.

* *Relative vorticity.* Vorticity resulting from both shear and curvature. This is typically used at the mesoscale level in the lower levels of the atmosphere.

* *Absolute vorticity.* Vorticity resulting from shear and curvature, plus the spin of the earth (planetary vorticity). In the Northern Hemisphere the earth adds positive (cyclonic) vorticity which increases from 0 radians/sec at the equator to 14 radians/sec at the pole (this is determined by $f = 2\Omega \cdot \sin(\phi)$). Most 500 mb vorticity charts use absolute vorticity.

6.7.7. Advection of vorticity. The vorticity equation directly relates vorticity advection to divergence and convergence.

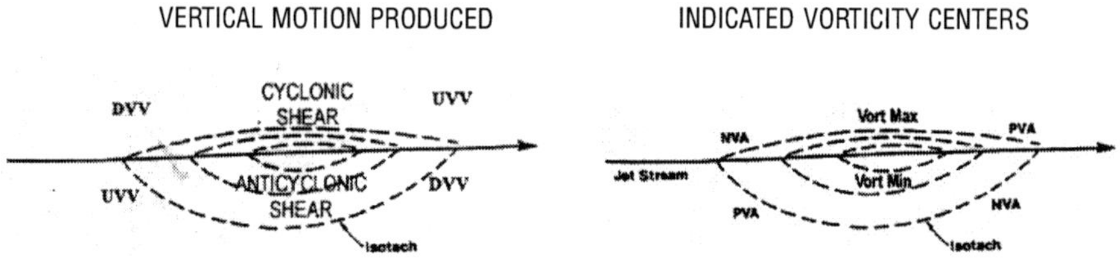

Figure 6-13a. Dynamics around a jet streak in straight-line flow. Isotachs are drawn as dashed lines. Upward vertical motion is suggested in the left front and right rear quadrants.

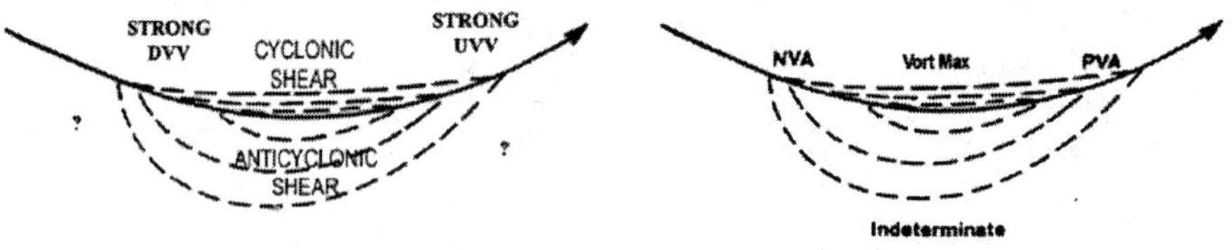

Figure 6-13b. Dynamics around a jet streak in cyclonically-curved flow. Cyclonic curvature is enhanced by cyclonic shear on the north side, but is negated by anticyclonic shear on the south side.

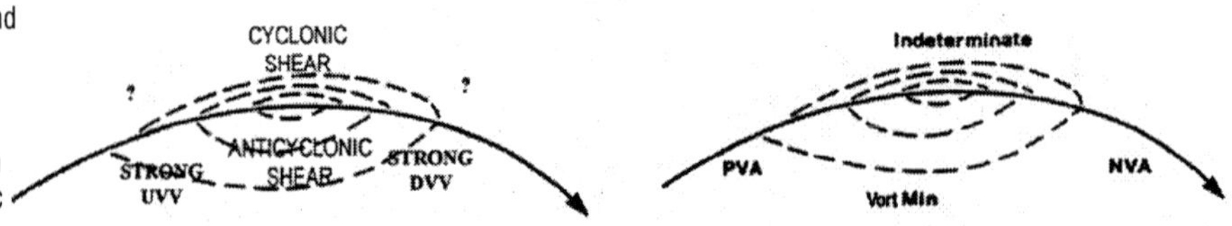

Figure 6-13c. Dynamics around a jet streak in anticyclonically-curved flow. Anticyclonic curvature is enhanced by anticyclonic shear on the south side, but is negated by cyclonic shear on the north side.

MOTION • 99

* *Positive vorticity advection (PVA)*. If 500 mb absolute vorticity is increasing with time at a given location, upper tropospheric divergence is indicated. This may imply upward vertical motion; if so, clouds and precipitation may result.

* *Negative vorticity advection (NVA)*. If 500 mb absolute vorticity is decreasing with time at a given location, upper tropospheric convergence is indicated. This may imply downward vertical motion; if so, clearing and fair weather may result.

6.7.8. **Vorticity and vertical motion**. Since PVA or NVA provides useful indicators of upper level divergence or convergence, it is often used as a method for locating areas of vertical motion.

* *Assumptions*. For PVA/NVA to work as indicators of vertical motion, the winds have to be nearly geostrophic. Unfortunately much of the significant weather that occurs is produced by winds that are out of geostrophic balance. Furthermore, it has to be assumed that the vorticity patterns are not moving faster than the upper-level winds, and that under the given circumstances 500 mb is in fact fairly close to the level of non-divergence. Additionally, the omega equation for vertical motion shows that PVA or NVA must increase with height for vertical motion to

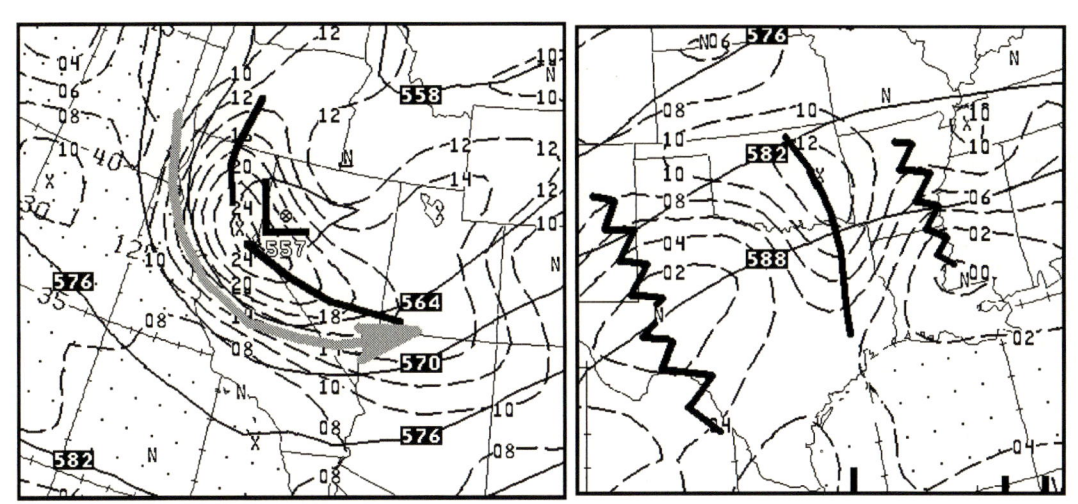

Figure 6-14a. Shear lobe. Note how there is very little crossing of the lobe (drawn in black) or the vorticity isopleths against the contour field. A shear lobe pattern tends to be generated by a jet streak or a jet axis (shown here). Jets streaks associated with shear lobes are sometimes called "channel jets" and the vorticity patterns are sometimes said to be "sheared". This is a cyclonically curved pattern, so PVA and upward motion can occur on the left front side of the pattern (see Figure X-XXb).

Figure 6-14b. Advection lobe. Note how the lobe (drawn in black) and vorticity isopleths sharply cross the contour field. Advection lobes are caused by short wave disturbances (in fact, they represent the short wave itself). PVA (and upward motion) is suggested ahead of the lobe axis. Note how the ridges on either side have been highlighted.

result — usually this kind of information is not readily available. The implications of this are that PVA and NVA should be used strictly as additional tools in understanding what is happening aloft.

Thermal advection. Under quasigeostrophic ("normal") conditions, PVA indicates divergence aloft. If this occurs, upward vertical motion can either be supported by low-level warm air advection or cancelled out by low-level cold air advection. Likewise, if NVA is present, downward vertical motion can either be supported by low-level cold air advection or cancelled out by low-level warm air advection.

PIVA/NIVA. It has been shown that PVA/NVA and thermal advection tend to cancel each other out. A new approach to the PVA/NVA problem has been to integrate both parts of the omega equation through evaluation of vorticity patterns by the thermal wind. This can be done by using the thickness contours rather than the 500 mb height contours to determine advection of vorticity. The advection of vorticity patterns through this technique is called PIVA and NIVA (positive and negative isothermal vorticity advection). Some informal studies have shown that using PIVA/NIVA has clear advantages.

Alternatives. There are other indicators of vertical motion available on computer systems that should be evaluated, such as convergence contours, isentropic analysis, and Q-vector analysis. Some of these will be covered elsewhere in this book. Furthermore, model forecasts do have output charts for the vertical motion field (even so, these should be weighted against model biases, model performance, mathematical limitations in the model, and other factors).

6.7.9. Vorticity and jet streaks. The convergent and divergent areas around jet streaks can be illustrated using vorticity principles. As we describe jet streaks, we will be using the left/right and front/rear coordinate systems, which refers to directions as if we are sitting on the jet stream or on a jet streak, straddling the axis and looking downstream. Therefore "left" is generally north of the jet axis, and "rear" generally refers to a westward direction (since our back faces west).

Jet streak in straight-line flow. See Figure 6-10a. All vorticity is produced by shear. The left side experiences cyclonic shear, so a vorticity max is produced. The right side experiences anticyclonic shear, so a vorticity min is produced. The right rear quadrant experiences PVA since it is behind the vorticity min, and thus upper divergence is indicated. The left rear quadrant experiences NVA since it is behind the vorticity max, and thus upper convergence is indicated. The left front quadrant experiences PVA since it is ahead of the vorticity max, and thus upper divergence is indicated. The right front quadrant experiences NVA since it is ahead of the vorticity min, and thus upper convergence is indicated.

* *Jet streak in cyclonic flow*. See Figure 6-10b. Vorticity is produced by both shear and curvature. The left side experiences both cyclonic shear and cyclonic curvature, so a strong vorticity max is produced. The right side experiences anticyclonic shear and cyclonic curvature, which negate each other. The right rear quadrant is indeterminate since there is no max or min ahead of it. The left rear quadrant experiences strong NVA since it is behind the strong vorticity max, and thus strong upper convergence is indicated. The left front quadrant experiences strong PVA since it is ahead of the strong vorticity max, and thus strong upper divergence is indicated. The left right quadrant is indeterminate since there is no max or min behind it.

* *Jet streak in anticyclonic flow*. See Figure 6-10c. Vorticity is produced by both shear and curvature. The left side experiences cyclonic shear and anticyclonic curvature, which negate each other. However the right side experiences anticyclonic shear and anticyclonic shear, which add up to strong anticyclonic vorticity. The right rear quadrant experiences strong PVA since it is behind a strong vorticity min, and thus strong upper divergence is indicated. The left rear quadrant is indeterminate since there is no max or min ahead of it. The left front quadrant is indeterminate since there is no max or min behind it. The right front quadrant experiences strong NVA since it is ahead of a strong vorticity min, and thus strong upper convergence is indicated.

<u>6.7.10. Shear lobes</u>. Shear lobes are elongated areas of vorticity which are aligned parallel to the wind flow.

* *Characteristics*. Shear lobes occur along the edges of a strong, elongated jet stream where curvature is broad and weak. Therefore shear and not curvature predominates in the vorticity equations.

* *Channel jets*. Segments of the jet stream which are associated with shear lobes are called channel jets. Vorticity advection may occur near the edges of a channel jet, but usually not along it, so dynamics tend to be weak. Channel jets tend to parallel the 500 mb contours and not cross them.

Jet axis. The axis of a channel jet lies lies along the area of strongest vorticity gradient and parallel to the 500 mb flow, with the positive vorticity on its left side and negative vorticity on its right side. The highest wind in the jet streak is between a couplet of the lowest and highest vorticity values.

6.7.11. Advection lobes.
Advection lobes are elongated areas of vorticity which are aligned perpendicular to the wind flow. These are usually what are referred to when a forecaster speaks of a "vorticity lobe" or "vort lobe".

* *Characteristics.* Advection lobes occur within a jet stream where curvature is somewhat sharp and shear along the edges of the jet stream is not well-defined. Therefore curvature and not shear predominates in the vorticity equations. Advection lobes generally indicate the possibility of a short wave trough or ridge at 500 or 700 mb and should be checked out further.

* *Advection jets.* Segments of the jet stream which are associated with advection lobes are called advection jets. Vorticity advection occurs along the jet axis, so dynamics tend to be strong. Advection jets tend to cross the 500 mb contours at an angle rather than parallel them.

* *Jet axis.* The axis of an advection jet is usually found in the area of strongest gradient between negative and positive vorticity, with positive vorticity to its left and negative vorticity to its right.

6.8. Q-VECTORS

6.8.1. Concept.
A Q-vector is a derived measure of vertical motion. It highlights areas where the thermal wind balance is being disrupted, which will cause compensating vertical motions. The Q-vector approach combines the two key contributions to vertical motion: temperature advection and convergence/divergence due to changes in vorticity advection with height. It is a good tool because vorticity advection, the indicator that has long been used by forecasters for decades, is often cancelled out by thermal advection. Likewise, vorticity advection can negate the anticipated effects of thermal advection. Q-vectors provide a quantitative way of measuring both elements. They cannot be easily computed by hand, and the output is quite difficult to find (especially on the Web), but certain software programs and weather analysis systems can easily calculate the values.

MOTION • 103

6.8.2. Usage. Q-vectors can be used at different levels to estimate the vertical motion at different heights in the atmosphere. Output fields at 500 mb and 300 mb tend to give the best results for a quick overview.

6.8.3. Patterns. Here are the four basic Q-vector patterns.

Q-vector pattern	Indicates...
Convergence	Rising motion at that level
Divergence	Sinking motion at that level
Points toward warm air	Frontogenesis
Points toward cold air	Frontolysis

6.9. ISENTROPIC ANALYSIS

6.9.1. Definition. Isentropic analysis refers to the diagnosis of vertical motion using isentropic surfaces (three-dimensional surfaces of equal potential temperature). Except for a brief period of use during the 1940s, the technique was mostly abandoned until the 1990s when better computing power made isentropic charts easy to make. They are frequently used for nowcasting during the winter season when diabatic effects (strong heating, etc) are less dominant.

6.9.2. Concept. Isentropic charts give forecasters an edge because in traditional constant-pressure maps (such as 500 mb charts), the parcels have a component of motion in or out of the page (up or down). However isentropic surfaces are three-dimensional, following parcel motion. If a car was a parcel of air under adiabatic conditions, an isentropic surface is like the ground. The ground coincides with the motion of the car. Except of course for the Dukes of Hazzard, whose car exemplifies diabatic flow. So instead of using several charts to determine where PVA and warm advection is the strongest, isentropic charts give this information on a single chart.

Figure 6-15. Isentropic surfaces bend upward over cold air masses and downward over warm air masses.

6.9.3. Distribution of potential temperature. Potential temperature (represented by θ or adiabats on the SKEW-T diagram) almost always increases with height — for example, it will be about 270 deg K near the ground, increasing to 320 deg K in the upper troposphere. Also, given a constant pressure (i.e. a constant

HORIZONTAL VIEW AT 850 MB

VERTICAL CROSS SECTION

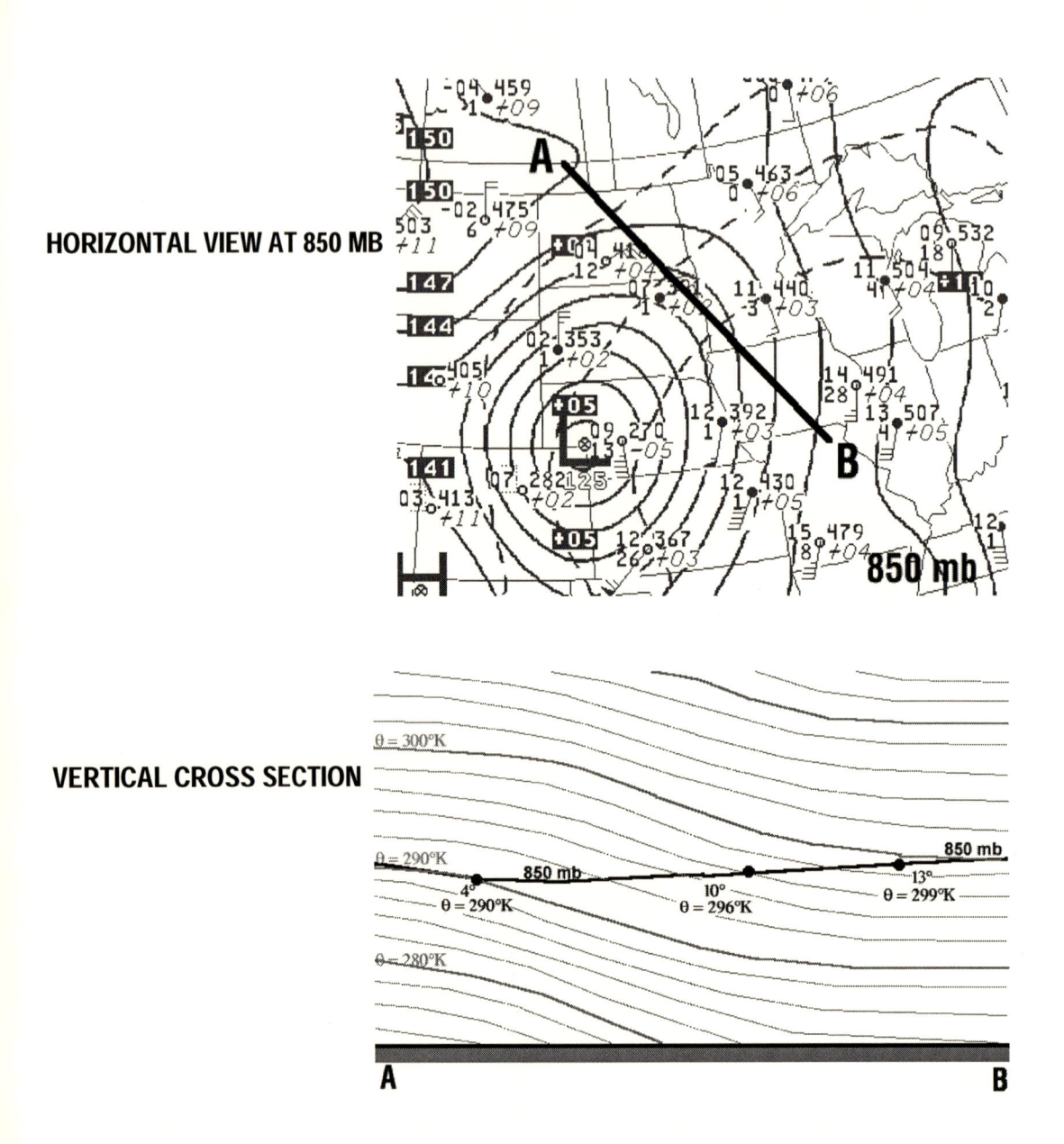

Figure 6-16. Examining the isentropic surfaces across a warm front in the north Plains. Because low-level temperatures cool as one goes north, the isentropic surfaces slope upward over the cold air. Air tends to cling to isentropic surfaces under adiabatic conditions. Therefore if wind blows in a direction that makes it rise upward along a surface, upward vertical motion is produced. The 850 mb chart shows that air is blowing predominantly from B to A (especially midway between the points), and the cross section shows that air flowing in this direction will rise along the isentropic surfaces, producing upward motion, clouds, and possibly precipitation.

altitude), potential temperature is lower in cold air than in warm air. This implies that the coldest potential temperatures are found near the ground in very cold air masses. The warmest potential temperatures are found in the stratosphere.

6.9.4. Isentropic surfaces. Isentropic surfaces are created by connecting points of equal potential temperature. These surfaces form a multiple-layered blanket, with the coldest surfaces near the ground and the warmest surfaces near the stratosphere (remember potential temperature increases with height). Furthermore, surfaces bend upward through colder air masses, similar to how blankets bend upward over your body when you're lying in bed. This is because the cold air masses have lower potential temperature and there must be room for colder surfaces contained within them. Isentropic surfaces are labelled in degrees Kelvin, so that a parcel at 1000 mb with a temperature of 32 deg C is on a 305 deg K surface.

6.9.5. Isentropic gradients. Where isentropic surfaces are vertically very close together, this indicates that the potential temperature is rising rapidly with height. In turn, this means that the air temperature is steady or rising with height, indicating an inversion. Often, this implies a frontal inversion.

6.9.6. Isentropic voids. Where isentropic surfaces are vertically very far from each other, it indicates that the potential temperature is rising weakly with height. In turn this means that the air temperature is falling rapidly with height, indicating an area of steep lapse rates (instability).

6.9.7. Motion. Air tends to cling to the potential temperature surface it is located at. This is because potential temperature is conserved; it can never be changed unless heat is added to or removed from the parcel. Therefore air tends to follow isentropic surfaces. Again, going back to the analogy of being in bed, an air parcel would be trapped between the sheets. Your body, representing a cold air mass, bends the surfaces upward. If the air is blowing from one side of the bed to the other, the parcel will be forced to rise as it moves toward the cold air mass. This is called "isentropic lift", and describes exactly what happens as the low-level flow moves from south to north across a warm front. Likewise, when the air parcel leaves the cold air mass, it descends to lower surfaces.

6.9.8. Vertical cross sections. One way of visualizing isentropic lift is by constructing cross-sections showing isentropes (lines of equal potential temperature). Of course, you have to assume that the winds are flowing along the plane of the cross section to make inferences about vertical motion, but still the cross section can be a useful tool.

6.9.9. Constant potential temperature maps. One popular way of seeing isentropic surfaces horizontally is through constant potential temperature maps. The variation in height of the isentropic surface is represented by height contours, and winds at the height of the isentropic surface are plotted. This shows the "topography" of the isentropic surface and readily indicates where vertical motion will occur. When winds are flowing from lower heights to higher heights, rising motion is indicated, and vice versa.

6.9.10. Finding an isentrope. To use constant potential temperature maps, you have to figure out which isentropic surface you want to look at. If it's too cold, you'll be too close to the ground and much of the chart will be unusable. If it's too warm, you'll be too high up and you may be missing what is happening in the lower and middle layers of the atmosphere. A rule of thumb is to locate the warmest surface temperature in your area of interest and select that as the isentrope. You can also look at cross-sections through your area of interest to select a good isentrope to look at. Usually in winter a good surface is 290-295 deg K, and in summer a good surface is 310-315 deg K.

6.10. CONDITIONAL SYMMETRIC INSTABILITY (CSI)

Conditional symmetric instability (CSI) is beyond the scope of this book as it's one of the most complicated processes to explain, however because of its importance it will be briefly addressed here. It produces an effect called slantwise convection. This normally consists of large bands of showery precipitation or weak thunderstorms oriented along the thickness isopleths (generally east-west).

Forecasters use cross sections perpendicular to deep-layer thickness isopleths which show momentum and equivalent potential temperature. CSI is found in near-saturated air where the slope of the momentum isopleths is steeper than the theta-e isopleths.

A simpler way to find areas of possible CSI is to look for deep layers where the atmosphere is nearly saturated, the temperature profile is nearly moist adiabatic, strong vertical speed shear is present, and a weak to moderate amount of synoptic-scale vertical lift is available. Subjective experience has found that these areas are most common in the entrance region to a jet in anticyclonically curved flow.

REVIEW QUESTIONS

1. Civilization has collapsed and a roving band of anarchists have appointed you as their weather sage. They are trying to decide whether to sack a warehouse tomorrow, and it's important that the weather be dry so that none of the goods get ruined. You noticed yesterday that the cirrus was moving west-to-east, and today it is moving west-northwest-to-east-southeast. What is your recommendation?

2. There is a long wave trough over the western U.S. and a ridge over the eastern U.S. There are currently 4 long waves around the northern hemisphere. Forecasters expect two long waves in Asia to merge, leaving 3 waves. Based on this scenario, what kind of weather can Chicago expect in the days ahead?

3. What are the two dynamic causes of vertical motion as identified in everyday synoptic analysis?

4. What are the two components of vorticity?

5. You are on a 747 flying across the Arctic, and are talking to the captain. He points out some unpredicted weather ahead: there is a storm system to the left and unusually clear skies to the right. Since you're up at 39,000 feet in the lower stratosphere, you'll be flying well above any of this weather. However the captain is concerned about fuel and asks you to help him find some cold, dense air, which will generate added lift and better range. Thankfully you remember about warm sinks and cold domes. Without making any assumptions about winds aloft, do you think the plane should bear left towards the storm, or right towards the unusually clear air?

6. One afternoon in February, Kansas City reports fair skies, a temperature of 34 deg F, a dewpoint of 29 deg F, and northwest winds at 15 mph. However upper-level charts show that it is under an area of very strong PVA (positive vorticity advection). A check on the computer shows that PVA even increases with height and the data is good. So why isn't the weather cloudy or rainy?

7. You are in Washington D.C. and are taking friends out for a picnic. The winds throughout the entire atmosphere are blowing south to north. CNN reports that a Senate hearing has generated a significant amount of hot air in the lower troposphere (but not enough for convection). Describe how the isentropic surfaces deflect over Washington D.C., and what direction you should head to avoid stratiform rain.

8. Which quadrants of a straight-line jet streak are associated with bad weather?

9. What type of Q-vector signature is associated with bad weather?

10. Why is important to use different isentropic surfaces depending on the season?

Hazy sunset in Emporia, Kansas under stagnant air mass. *(Tim Vasquez)*

7 BAROTROPIC SYSTEMS

I put my air conditioner in backwards. It got cold outside. The weatherman on TV was confused. "It was supposed to be hot today!"

STEVEN WRIGHT, 1984

Barotropic weather systems contain no advection, so vertical motions derived from the omega equation include nothing from the advection term. However, if there is increasing positive vorticity with height, upward vertical motion will result.

7.1. COLD CORE BAROTROPIC LOW

A cold-core barotropic low is comprised of cold air throughout the atmosphere. It causes low thicknesses within the air, resulting in low upper-level heights.

7.1.1. Occlusion. The most common example of a cold-core barotropic low results from the occlusion of a baroclinic system. The occlusion usually fills due to boundary layer convergence. The cyclonic circulation strengthens with increasing height. The system is vertically stacked, but it may have a small amount of vertical tilt (i.e. an imaginary line drawn between the surface low and the 500 mb low will usually be vertical, but may tilt slightly from vertical). Since the coldest air is in the center of the low, the system is not directly associated with fronts. The occluded low is always the result of a baroclinic system which has occluded (the low has completely wrapped air masses into itself, which mix and become homogenized). It occurs north of the polar front jet. The low usually weakens or absorbs into a larger semipermanent low pressure system, such as those found in the northern Atlantic or northern Pacific. While the occluded low is dissipating, fresh air mass contrasts to its south often lead to the development of new baroclinic system (new waves).

7.1.2. Cutoff Low. These are isolated upper-level lows that are found south of the polar front jet. They usually develop along the Pacific coast when a strong baroclinic high off of British Columbia transports cold polar air southward off the coast of Washington and California. The polar front jet stays to the north, causing the circulation to isolate itself and close off. The circulation is usually weak and difficult to find at the surface. Cutoff lows typically stay in the northwest Mexico region for days, then wander towards Texas or the Gulf of Mexico, where they may interact with rich moisture and produce thunderstorms. In rare cases a cutoff low may move westward into the Pacific and disappear. Most computer forecast models handle cutoff lows poorly; the most reliable movement indicator is Henry's Rule, which states that a strong short wave along the main polar front jet must come within 1000 miles of the cutoff low; the circulation will then open up.

7.2. WARM CORE BAROTROPIC LOW

A warm-core barotropic low is comprised of warm air throughout the atmosphere. This warm air causes higher thicknesses within the air, resulting in high upper-level heights. The low density of the warm air also causes low pressure at the surface. The system is barotropic, so it is vertically stacked, though it may have a slight amount of tilt. The cyclonic circulation weakens with increasing height, and may become a high aloft. Since the warmest air is in the center, no fronts are directly associated with the system. The low occurs south of the polar front jet, and is caused by low-level heating of the air mass. Any precipitation is relatively symmetrical around the center.

7.2.1. Heat (Thermal) Low. The most obvious example of a warm-core barotropic low is the heat low, usually seen over the desert southwest during the summer months (a surface trough may extend into the San Joaquin Valley of California). Its energy source is diabatic, originating from strong solar heating on dry desert ground. The exact center of the surface low may be ambiguous and difficult to find, due to uneven heating. Diurnal thunderstorm activity may develop if there is sufficient moisture.

7.2.2. Tropical Cyclone. Another example of a warm-core barotropic low is the tropical cyclone, which is related to the heat low but is more of an adiabatic system. It does receive substantial heat energy from the ocean surface, but the moisture present in the air mass and strong upward vertical motions cause extensive adiabatic warming due to the release of latent heat within the weather system. The circulation weakens only gradually with height, and it is sometimes necessary to go up to 200 mb to find any anticyclonic circulation.

7.3. COLD-CORE BAROTROPIC HIGH

The cold-core barotropic high contains cold air at the center of the high. The high density causes high surface pressures, while the low thicknesses result in low heights aloft. Since the system is barotropic, the system is vertically stacked, but may have a slight amount of tilt. The circulation weakens with increasing height. Since the coldest air is in the center, it is not directly associated with fronts. The high is usually north of the polar front jet.

7.3.1. Polar air source. The most common example of a cold-core barotropic high is a polar air mass in its source region. It forms due to intense cooling of the air in the low levels. As the air mass moves south, temperature contrasts in the air mass begin developing, causing the high to become a baroclinic high.

7.4. WARM-CORE BAROTROPIC HIGH

The warm-core barotropic high contains a substantial amount of warm air throughout the atmosphere. Since it is barotropic, it is vertically stacked (but may have some tilt). The circulation strengthens with height. Since the warmest air is in the center of the system, it is not associated with fronts.

7.4.1. Subtropical High. The subtropical high is a semipermanent feature in the subtropics. It is usually strong at the surface and aloft. The high is caused by air moving northward aloft from the tropics. It stops moving north at about 30 deg latitude due to the Coriolis effect and begins accumulating. This convergence causes an increase in mass and the air sinks (subsides). The surface flow moving northward from the subtropical high is referred to as the prevailing westerlies, while the flow moving southward is called the trade wind.

7.4.2. Cutoff High. The cutoff high is a warm barotropic high located north of the polar front jet (either the main one or a southern branch of one). The surface high is typically weak. It forms when a warm air mass aloft is transported to a high latitude by a strong southerly flow. The flow later becomes more zonal, cutting off the upper-level high, leaving a warm pool to the north. Cutoff highs are rare, but are seen more often in the Atlantic regions. When they do develop, they may cause an omega block pattern.

7.4.3. Plateau High. The plateau high is very similar to the subtropical high, however it forms due to radiational cooling.

REVIEW QUESTIONS

Given each situation below, name the type of barotropic system that is present as well as its thermal structure (e.g. cold-core barotropic high).

1. In Glasgow, Scotland in March there is a week-long spell of rainy weather with occasional storms. A pilot reports very strong winds aloft. The jet stream is well to the south.

2. In Norfolk, Virginia in August, a sweltering day gradually deteriorates by evening into a prolonged, raging storm with 60 mph winds, driving rain, and thunder.

3. In Minneapolis, Minnesota in February it is sunny and calm with temperatures rising as high as the mid-60s. Meanwhile all of the southern United States is experiencing rainy, unsettled weather due to the jet stream traversing east-west across that area.

4. In Yellowknife, Northwest Territories in January it is clear and calm with a constant temperature of -35 deg F. There is about 6 inches of snow on the ground.

5. In Berlin, Germany in December it is fair and very warm. A strong jet stream is well to the south.

6. In Albuquerque, New Mexico in January it is rainy, cold, and unsettled. A jet stream is well to the north.

7. In Lousiana in June there is a broad surface trough with scattered thunderstorms. Winds aloft are very weak.

8. In Charleston, South Carolina in September it is hazy, clear, and calm, with highs in the 90s.

9. In Yakutsk, Russia in September there have been several nights of strong radiational cooling, with lows dipping to near 20 deg F. Winds have been very light.

10. In Baltimore, Maryland in June it is humid and sunny, and a drought continues. Winds aloft are light.

BAROTROPIC SYSTEMS • 115

COLD-CORE BAROTROPIC LOW — Cutoff low

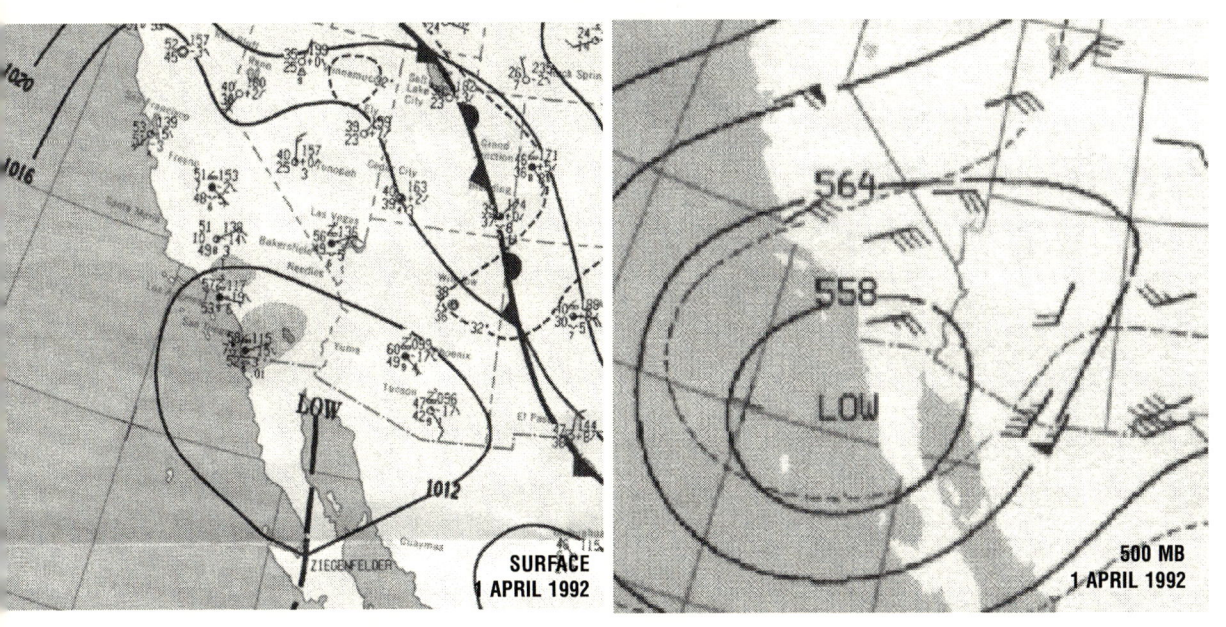

Figure 7-1. COLD-CORE BAROTROPIC LOW (CUTOFF LOW). The cutoff low becomes stronger with height. It is sometimes not apparent on the surface except for weak pressure falls and scattered shower activity.

COLD-CORE BAROTROPIC LOW — Occlusion

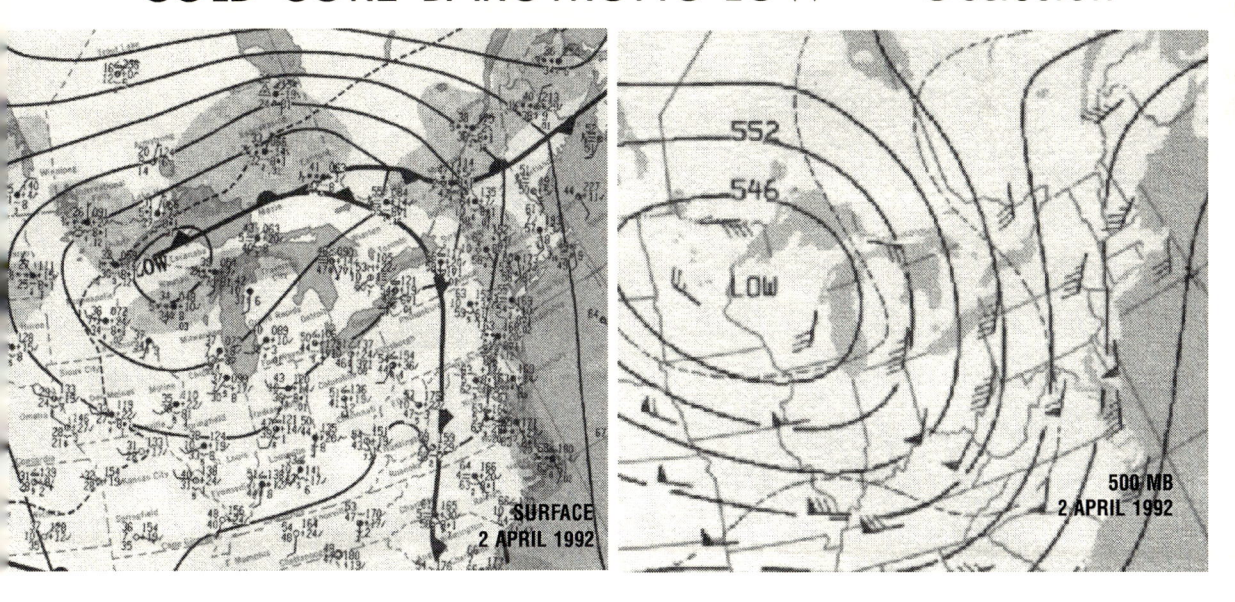

Figure 7-2. COLD-CORE BAROTROPIC LOW (DECAYING WAVE). Also known as an occlusion, this is a form of a barotropic cold-core low.

WARM-CORE BAROTROPIC LOW — Heat low

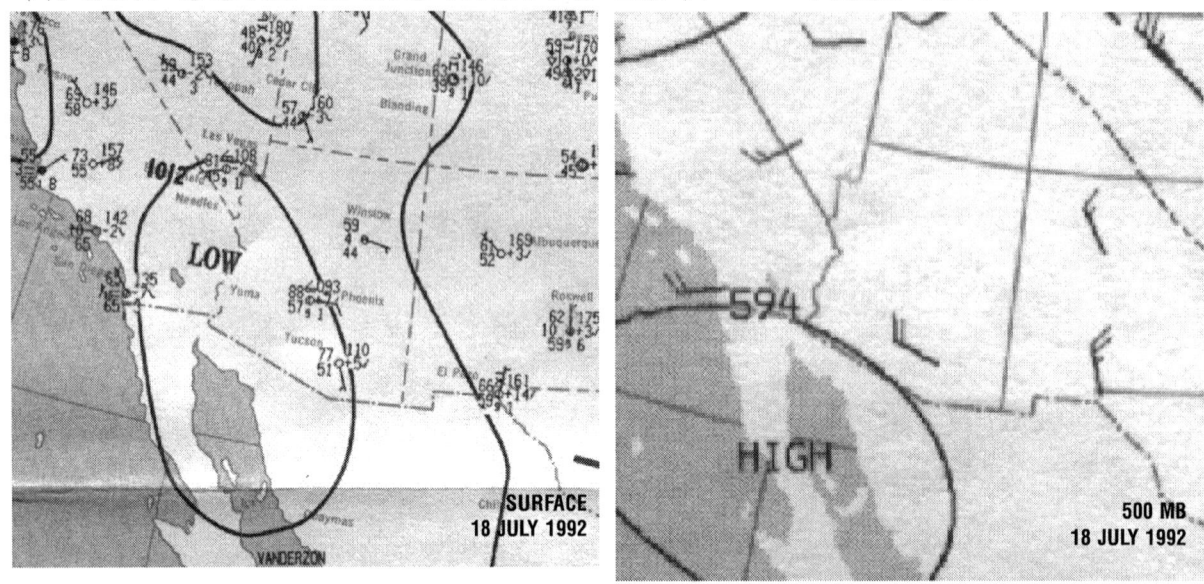

Figure 7-3. WARM-CORE BAROTROPIC LOW (HEAT LOW). Heat (thermal) low, which is a warm-core barotropic low. It is structurally related to the hurricane but gets its energy from insolation, rather than from the release of latent heat.

WARM-CORE BAROTROPIC LOW — Tropical cyclone

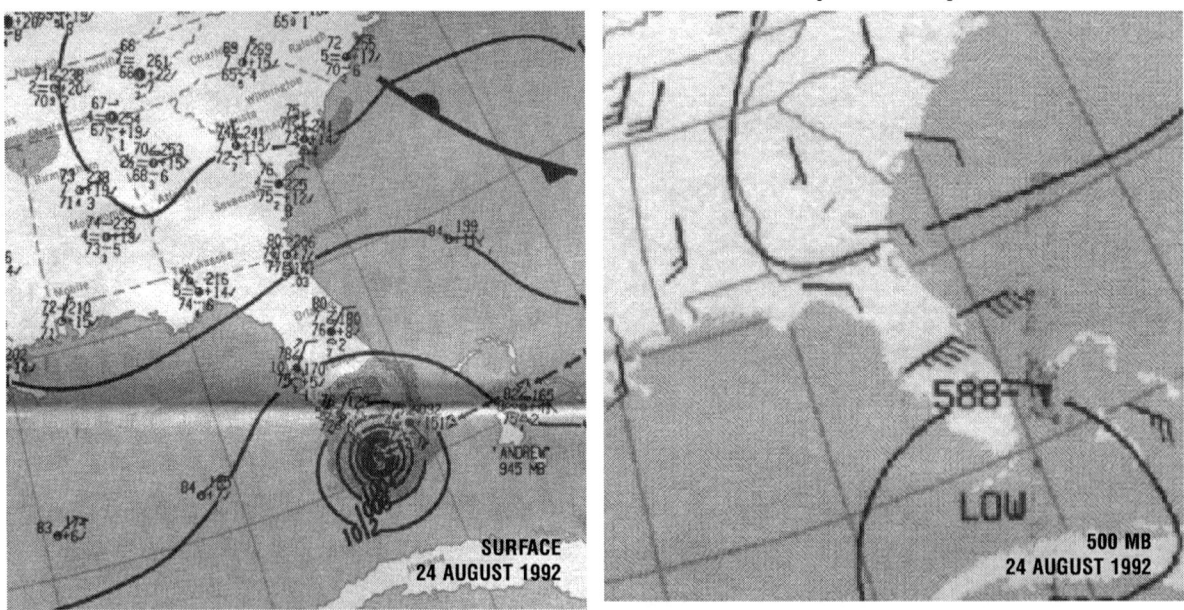

Figure 7-4. WARM-CORE BAROTROPIC LOW (TROPICAL CYCLONE). This type of feature is a warm-core barotropic low, which implies that high heights should be found aloft. The 500 mb chart, however, shows a low. It is frequently necessary to go to 300 or 200 mb to find the upper-level high over the storm. In this case, the storm is Hurricane Andrew.

BAROTROPIC SYSTEMS • 117

COLD-CORE BAROTROPIC HIGH — Polar source region

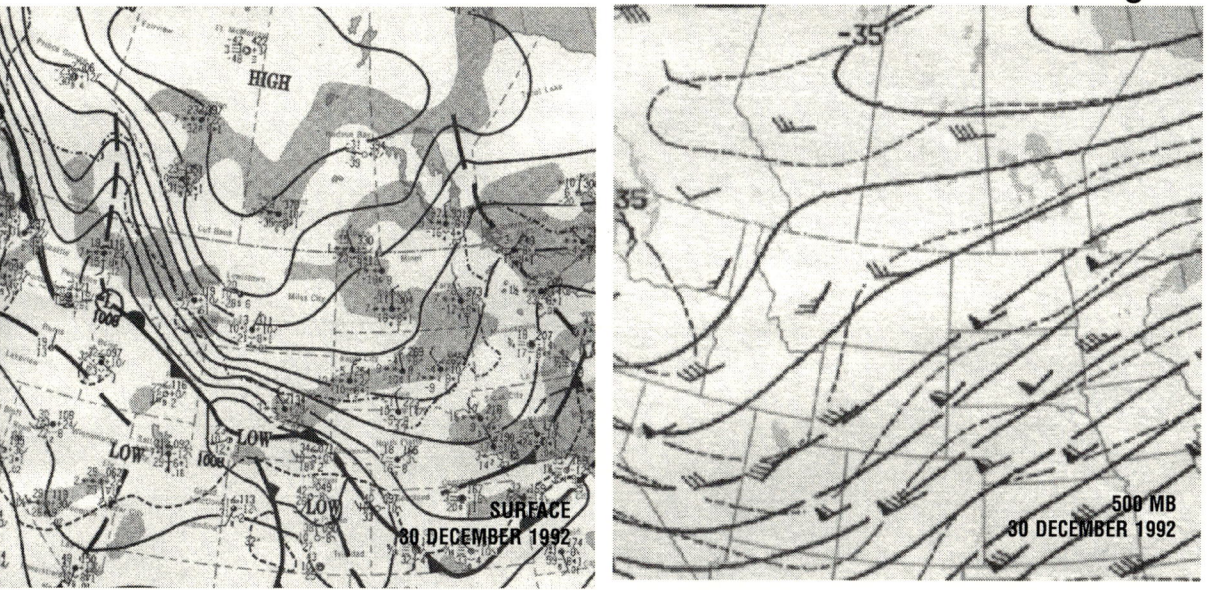

Figure 7-5. COLD-CORE BAROTROPIC HIGH (POLAR AIR MASS SOURCE). Cold-core barotropic high over northern Saskatchewan. Notice how it sits directly under a sharp trough aloft (low pressure). This high is continental polar air sitting in its source region over snow-covered areas of Canada. Two days later it evolved into a baroclinic high and shifted southward.

WARM-CORE BAROTROPIC HIGH — Cutoff High

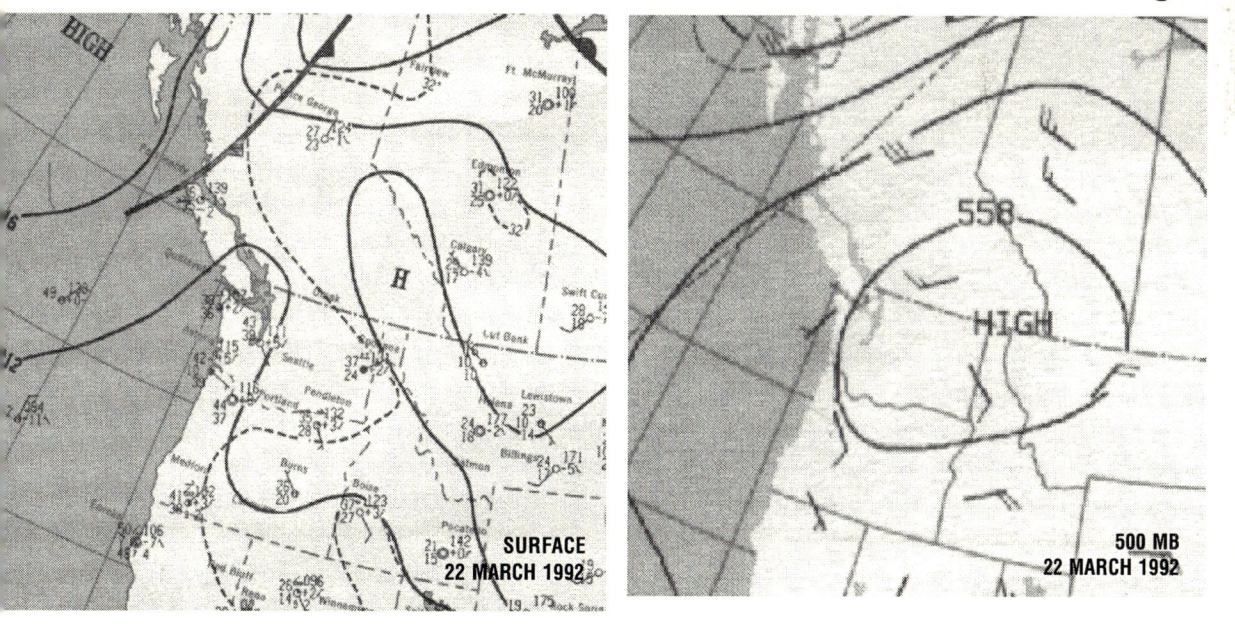

Figure 7-6. WARM-CORE BAROTROPIC HIGH (CUTOFF HIGH). A cutoff high, a form of warm-core barotropic high. It is very similar to a subtropical high, but is quite rare.

118 • WEATHER FORECASTING HANDBOOK

WARM-CORE BAROTROPIC HIGH — Cutoff High

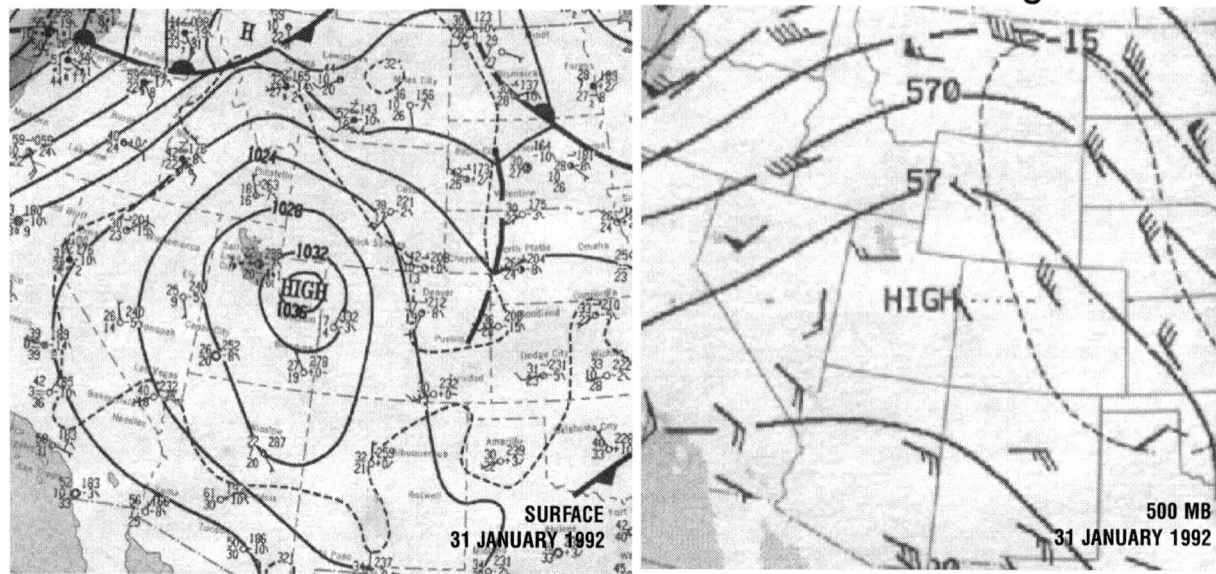

Figure 7-7. **WARM-CORE BAROTROPIC HIGH (PLATEAU HIGH).** Plateau high over Utah. There is little temperature advection around it, and an upper-level ridge sits above it. In Salt Lake City, persistent fog was accompanied by daytime highs in the 30s and lows around 20.

WARM-CORE BAROTROPIC HIGH - Subtropical ridge

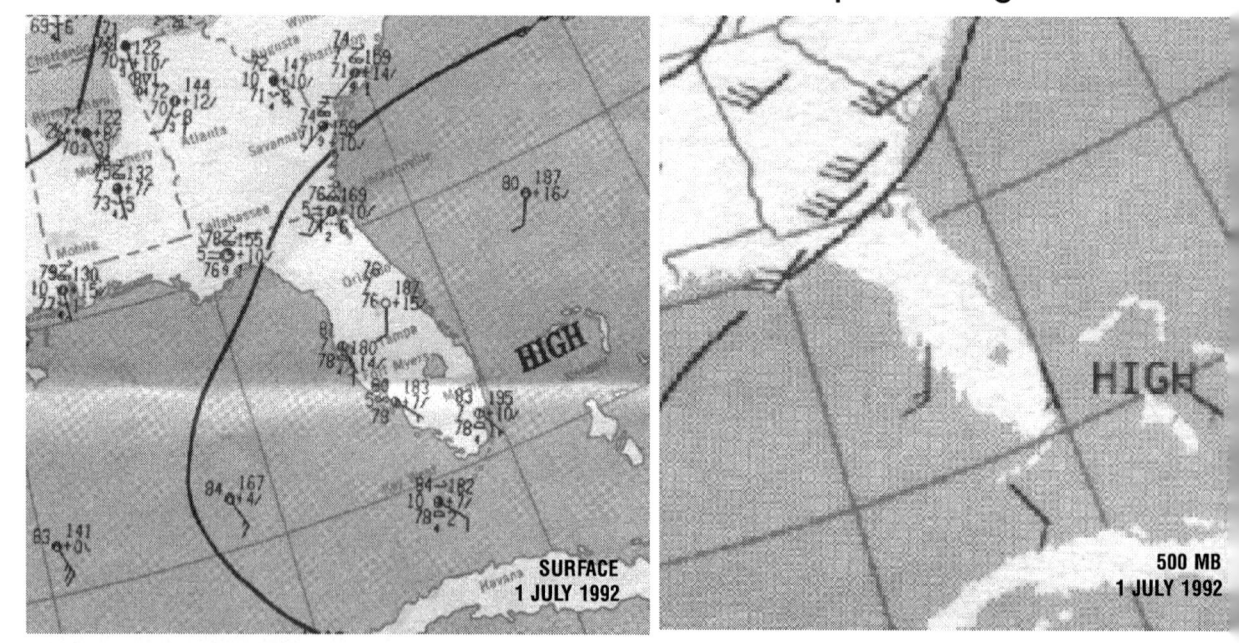

Figure 7-8. **WARM-CORE BAROTROPIC HIGH (SUBTROPICAL HIGH).** Warm-core barotropic high over the Bahamas. It translates to a high directly aloft. This is a classic example of a subtropical high, which usually serves as a source for maritime tropical air.

Vaulted cloud layers ahead of squall line near Stroud, Oklahoma. *(Tim Vasquez)*

8 BAROCLINIC SYSTEMS

The wind in the wires made a tattle-tale sound
And a wave tumbled over the railing
And every man knew, as the captain did too,
T'was the witch of November come stealing.
The dawn came late and the breakfast had to wait
When the gales of November came slashing
When afternoon came it was freezing rain
In the face of a hurricane west wind

GORDON LIGHTFOOT
Wreck of the Edmund Fitzgerald, 1976

The baroclinic weather system is driven by baroclinic instability. Baroclinic instability describes the process which results from the difference in temperature between the polar and equatorial latitudes. The baroclinic weather system is the mechanism that relieves this strong gradient, acting to convert the potential energy stored by the thermal gradient into kinetic energy that drives the baroclinic system's circulation. When enough kinetic energy is depleted, the system dissipates, and the cold and warm air masses have been distributed and mixed along the gradient. So in effect the gradient has been relieved. Naturally, all baroclinic weather systems have temperature advection as their primary feature.

8.1. BAROCLINIC LOW

<u>8.1.1. Development</u>. A baroclinic low, also known as an extratropical cyclone or a frontal low, forms along a stagnant polar front boundary. Slight upper-level divergence moving directly over the boundary causes pressures to fall along the front, which in turn causes the air masses from each side of thefront to advect towards the low pressure area to fill it in. These air masses are deflected to the right due to the Coriolis effect, resulting in a counterclockwise circulation around the developing wave. Cold air advection southward induces height falls, which in turn increases upper-level vorticity, which in turn increases upper-level divergence, which in turn deepens the surface low. This "chain reaction", known as self-development, is what causes frontal lows to sometimes become very deep.

<u>8.1.2. Dissipation</u>. The low begins filling when the air masses involved in the circulation wrap around the low (occlusion). This mixes out the temperature contrasts, turning the baroclinic low into a cold-core barotropic low. This takes away the upper-level support, breaking the chain reaction. The occluding low then begins filling and eventually dissipates or absorbs itself into nearby semipermanent low pressure areas. However, temperature contrasts remain to the south of the low, so a new one often forms to the south and may follow a similar life cycle as the original low.

<u>8.1.3. Examples</u>

8.1.3.1. Great Plains low. A deep area of low pressure is in the northwest United States, and the jet flows from California into the Great Plains. A lee-side trough begins forming just to the east of the Rocky Mountains due to subsident warming and vortex stretching (an orographic effect).

8.1.3.2. Bombs. Explosive cyclogenesis within a baroclinic low is often referred to as a "bomb". The favored location for these events are in the western Atlantic off the coast of New England during the winter months. Explosive cyclogenesis is considered to be a deepening of about 18 mb in 24 hours.

8.2. BAROCLINIC HIGH

A baroclinic high is in effect a "frontal high". A high pressure area cannot support a front (the diverging surface air automatically weakens thermal contrasts), but baroclinic highs represent the source of air masses involved in frontal systems.

8.2.1. Development. The classic life cycle of a baroclinic high starts with it developing in a baroclinic zone within the cold air mass. It intensifies beneath the upper-level convergence ahead of a short wave ridge and moves in an easterly or southeasterly direction.

8.2.2. Dissipation. Eventually, the air at the center warms due to subsidence and heating from below (adiabatic and diabatic warming), and the high becomes barotropic, being absorbed into the subtropical ridge, often becoming part of the Bermuda high.

8.2.3. Movement. Baroclinic high movement can be estimated by taking 50% of the 500 mb flow or 70% of the 700 mb flow in the early development stages.

8.2.4. Intensification. Further development is favored when the short-wave ridge remains within 300 to 450 nm upstream. They may also strengthen under confluent flow in the jet stream pattern. When height rises increase over the surface high, this indicates that self-development is underway. Surface highs north of the jet or north of the tightest 1000-500 mb thickness gradient are favored for further development.

8.2.5 Braking. Braking mechanisms slow down and eventually stop the development of baroclinic highs. Although self-development causes baroclinic highs to strengthen, the intensification initiates other processes which slow down and ultimately stop the development. Braking mechanisms for highs are much more efficient for highs than for lows, so highs rarely get as strong as low pressure circulations.

8.2.5.1. Boundary layer processes. As a baroclinic high builds due to convergence aloft, the low-level anticyclonic circulation increases. Anticyclonically-curved flow in the boundary layer causes low-level divergence, which will partially offset the mass being added by the system aloft. Friction slows down surface air parcels, which reduces the Coriolis effect and causes the pressure gradient force to be dominant; the winds diverge out of the high more strongly.

8.2.5.2. Adiabatic temperature changes. Subsidence is a warming process. To force this air to sink takes energy out of the high, reducing the energy available for development. It often limits the intensity that highs can attain.

8.2.5.3. Indicators of braking. When the 1000-500 mb thickness ribbon spreads apart within the high with time (a weakening thermal gradient), low-level divergence is predominating and the high is weakening. Another indicator of weakening is when the surface high center is located on the south side of the jet and has moved to a higher 1000-500 mb thickness line; adiabatic warming is indicated. Finally, an indicator of weakening is when upper-level height rises are diminishing; upper-level support is decreasing.

REVIEW QUESTIONS

1. Describe the chain reaction of self-development with a baroclinic low.

2. Are coastal regions more favorable for baroclinic development during the summer or during the winter?

3. South of a baroclinic low, a subtropical jet begins producing extensive rain above the warm sector. The warm sector region has been very dry. How will this affect the baroclinic low?

4. A large low pressure area is located over Missouri. A strong upper level system has arrived and is causing intense deepening of the low. What will happen to the low at this point?

5. How can you use thickness charts to evalute where baroclinic low development is most likely?

6. In January, a low pressure system moves off the New Jersey coast into the Atlantic Ocean, where it begins rapidly deepening. What is the term for this type of development?

7. What is the final stage in the life cycle of a baroclinic low?

8. How is a baroclinic high different from a barotropic high?

9. Is a baroclinic high most likely to dissipate quickly if it moves into mountainous areas or flat areas? Why?

10. How can you use thickness charts to evaluate how quickly a baroclinic high is dissipating?

BAROCLINIC SYSTEMS • 125

BAROCLINIC LOW

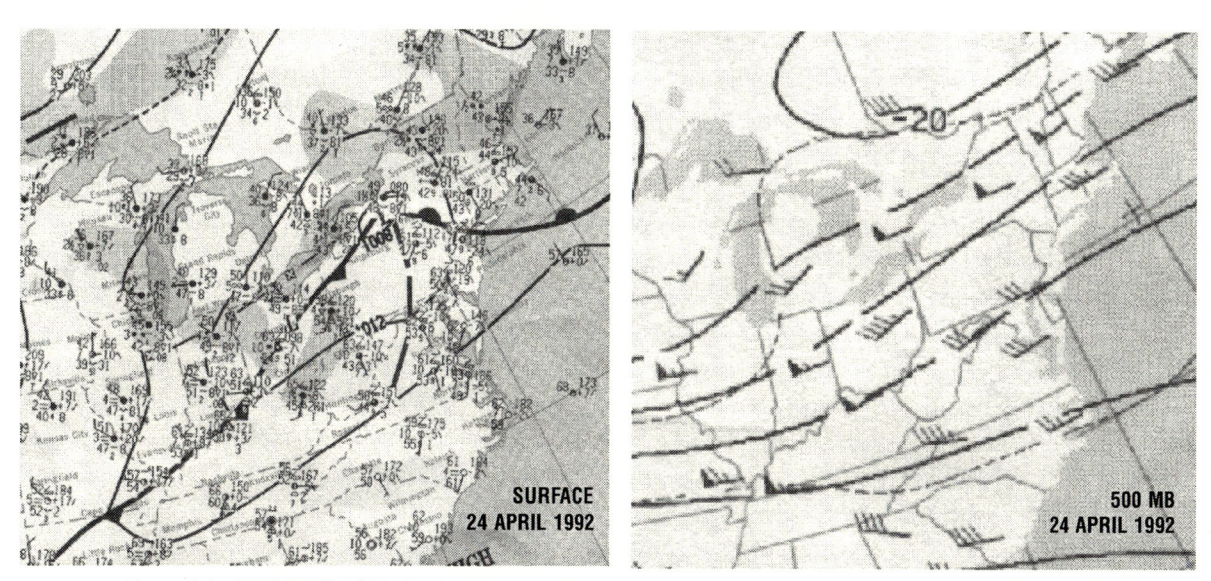

Figure 8-1. BAROCLINIC LOW. Two baroclinic lows, one over western New York and another over Ohio. The polar front jet sits directly on top of them, giving them dynamic support. Meanwhile, the cold and warm air masses converging into the lows help support the polar front jet.

BAROCLINIC HIGH

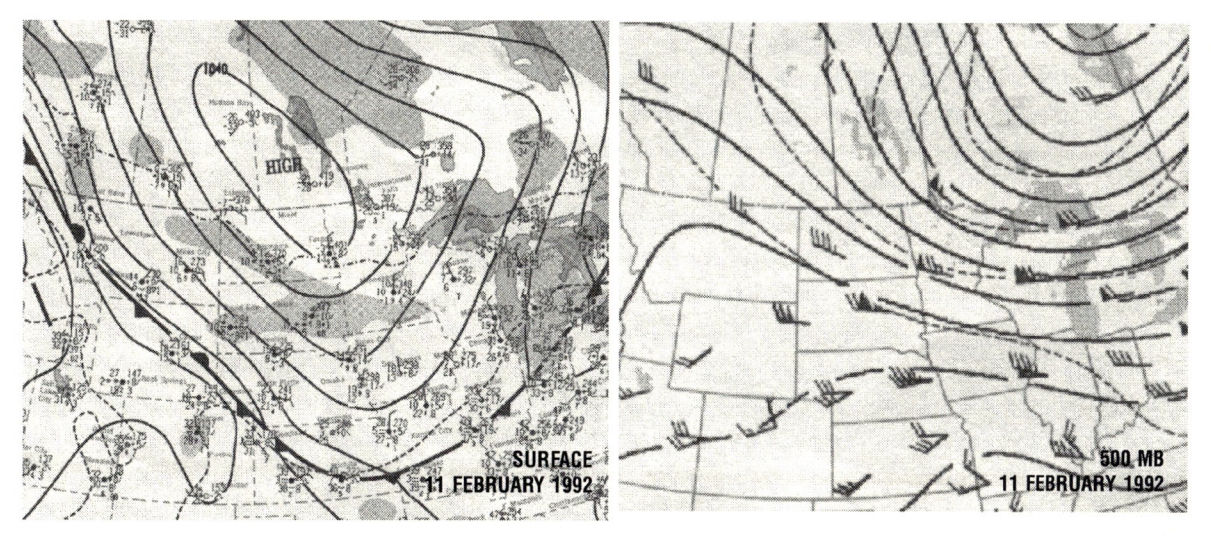

Figure 8-2. BAROCLINIC HIGH. Baroclinic high over southern Manitoba. Directly above it, the 500 mb chart shows a 50 kt jet in place.

Low-precipitation supercell in Norman, Oklahoma. *(Tim Vasquez)*

9 CONVECTIVE WEATHER

The wind began to rock the grass with threatening tunes and low,
He flung a menace at the earth, a menace at the sky.

The leaves unhooked themselves from trees and started all abroad;
The dust did scoop itself like hands and throw away the road.
The wagons quickened on the streets; the thunder hurried slow;
The lightning showed a yellow beak, and then a livid claw.

The birds put up the bars to nests, the cattle fled to barns;
There came one drop of giant rain, and then, as if the hands
That held the dams had parted hold, the waters wrecked the sky;
But overlooked my father's house, just quartering a tree.

EMILY DICKINSON
A Thunderstorm, 2001

When the weather becomes sultry and oppressive and giant towers billow into the sky, the forecast⋯ be familiar with convective forecasting princ⋯ The foundation of this knowledge is the three key prerequisites for convection: **moisture**, a moist layer of sufficient depth in the lower or middle troposphere; **instability**, with a steep lapse rate within or above the moist layer to allow for a substantial positive lift area; and **lift**, with enough lifting of a parcel from the moist layer to allow it to reach its level of free convection (LFC). If all three of these conditions are met, the chance of convection is excellent. Greater amounts of moisture, instability, and lift will almost guarantee the formation of thunderstorm activity.

9.1. THUNDERSTORM STRUCTURE

A thunderstorm is made up of two parts: the updraft and the downdraft. They are distinct features which are caused by different processes and are responsible for different weather phenomena.

9.1.1. Updraft. The updraft is what fuels the storm. It consists of warm, moist air originating from storm **inflow** that is buoyant and rises rapidly. This frequently takes the appearance of a rapidly-bulding cumulonimbus tower. Exceptionally strong updrafts can take on rotation, sometimes producing a tornado under the base of the tower. As the tower grows, precipitation particles begin forming in the top of the tower, and eventually they fall out of the storm. This leads to the next part of the storm . . .

9.1.2. Downdraft. This feature is caused by precipitation particles accumulating at the top of the storm. The particles coalesce while others evaporate (chilling the air and causing it to sink); downward motion quickly begins occurring, forming the downdraft. It can fall directly back into the updraft, or in stronger storms, just downwind of the updraft (to the northeast). The downdraft, since it is sinking and causes cloud droplets to evaporate, appears visually as clear, bright air but may be filled (and darkened) with rain and hail. As the downdraft reaches the ground and spreads horizontally, it is known as **outflow**. The leading edge of the outflow is called the **gust front**. The downdraft is responsible for the vast majority of the heavy rain, hail, and high winds in a typical storm. It is important to note that the downdraft has a strong tendency to choke out the updraft by falling directly onto it or undercutting it, which ends the life cycle of the storm. However, as we shall

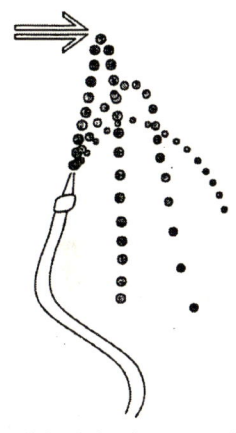

Figure 9-1. A thunderstorm updraft can be likened to the upward stream from a garden hose. The downdraft can be likened to the falling cascade of water droplets. If the upper-level winds are stronger than the low-level winds, the cascade (downdraft) will be carried downwind and will not "clog" the updraft. Therefore strong wind shear (difference in winds with height) supports long-lived thunderstorms. (NOAA/NWS)

Roman lightning protection
Among the Romans seal-skins were considered as an infallible preservative against lightning. Timorous persons anxiously crept into a tent covered with seal-skins when a storm was impending; and the Emporor Augustus, who seems to have been particularly afraid of thunder, always had a seal-skin near at hand. Arago relates that in the Cevenne Mountains, where Roman colonies were founded, the peasants carefully collect the skins of serpents, which they wear round their hats, and armed with this talisman, fearlessly brave the storm. Probably, these serpent-skins replaced among the people the more rare and costly seal-skins, which could only be purchased by the rich.

DR G. HARTWIG,
"The Aerial World," 1886

Figure 9-2. Yukon thunderstorm. Convective weather is not simply confined to warm climates. This thunderstorm was photographed along the Alaska Highway near Snag, Yukon on 5/27/99. Snag also holds the record for the all-time lowest official temperature in North America, at minus 83 deg F.

see, stronger storms circumvent this by handing off the updraft action to a different, newer updraft or by relying on strong upper-level winds to push the downdraft away from the updraft.

9.2. THUNDERSTORM TYPES

Although there are various types of thunderstorms, they are actually seen as part of a continuous spectrum. Therefore, a squall line can have characteristics of a multicell line, and so forth. In the examples below, we will focus on "ideal candidates" from each storm type.

9.2.1. Unicell. The simplest type of storm is the basic unicell (single-cell) thunderstorm, described by classic textbook thunderstorm diagrams. It is often regarded as the summertime popcorn-type airmass storm, though a pure unicell is rare since the atmosphere tends to produce multiple updraft cells. If a unicell develops in moderately unstable conditions, the pulse severe storm occurs. However, severe weather is usually confined to small hail and brief gusty winds. Both the unicell and pulse storm have short lives (less than an hour), since the lack of multiple updrafts precludes any new updraft from prolonging the storm.

9.2.2. Squall Line. One of the more common, simple storms you will see is a squall line, a long unbroken line of thunderstorms which extends for hundreds of miles in length. It is common ahead of strong fronts and troughs. The cloud form is very linear and organized. A classic squall line has its updraft on the east (front) side, with the downdraft on the west (back) side. This unique orientation and the linear nature of the gust front acting as a front-like wedge tends to push the updraft along rather than cutting it off, which means the updrafts are sustained for many hours and resulting in a steady-state storm. Squall lines may rage for days unless they move into an area of less moisture, unfavorable upper-level wind conditions, or higher stability. Squall lines have a high amount of cell competition that is, there are so many cells along the line that it's difficult for any one cell to take over and become severe. This is why tornadoes are rare and short-lived, and hail (although common) remains rather small. However, the forward motion combined with favorable downdraft intensity can cause strong straight-line winds as the storm marches eastward. The most common mode of small-scale severe weather formation is when a squall line breaks into segments. The cell to the north of the break gets an extra shot of fuel and can become temporarily dominant.

9.2.3. Multicell Cluster. The most common type of storm is the multicell thunderstorm, which encompasses a broad spectrum. The cluster variety frequently appears as larger thunderstorms with many different updraft and downdraft cells in different stages of formation and decay. The cluster usually starts as a unicell, but as downdraft air spreads into outflow, new updraft cells are generated along its periphery. More outflow develops, leading to more updraft cells, and all this seemingly takes place in a random pattern, with anywhere between two and fifty distinct cells taking place at once in an area the side of a city to a small state. The weaker multicell cluster usually forms the vast majority of summertime air-mass thunderstorms. A weak multicell cluster on radar plots usually looks like a chaotic area of moderate-intensity cells, and cell competition is noticeably high.

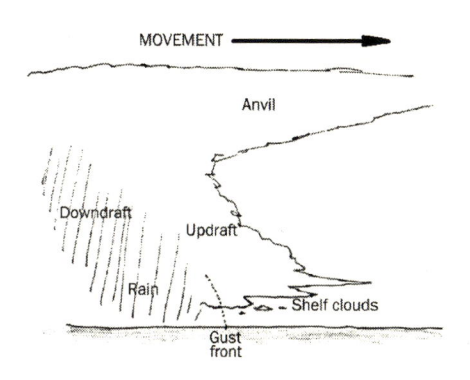

Figure 9-3. Two-dimensional slice through a typical squall line. The updraft tends to orient itself on the leading edge of the storm, with a trailing area of rain. Some small hail may occur along the leading edge of the rain. Straight-line hodographs with strong bulk shear tend to favor squall line scenarios.

9.2.4. Multicell Line. When instability is high, solar heating is present, and vertical wind shear is strong, a multicell line may form. These storms do not have random development along gust fronts. Rather, activity concentrates along a very short line about 10-30 miles long, where the old cells die out on the north part of the line, and newer cells sprout up on the southern part. The older cells turn into rainy downdrafts, while the newer cells are composed of towering cumulus clouds. This line of towering cumulus comprises a flanking line. As the most active cells die and the flanking line towers mature, an apparent cycle of "growth down the flanking line" is seen. This process, called backbuilding, tends to propagate the storm eastward rather than northeast. The majority of springtime severe storms are multicell lines. They can produce weak tornadoes, very large hail, torrential rains, and strong winds.

9.2.5. Supercell. This type of storm, which is a storm that has achieved a steady-state updraft-downdraft life cycle and often contains a rotating updraft, will be covered separately in the next section.

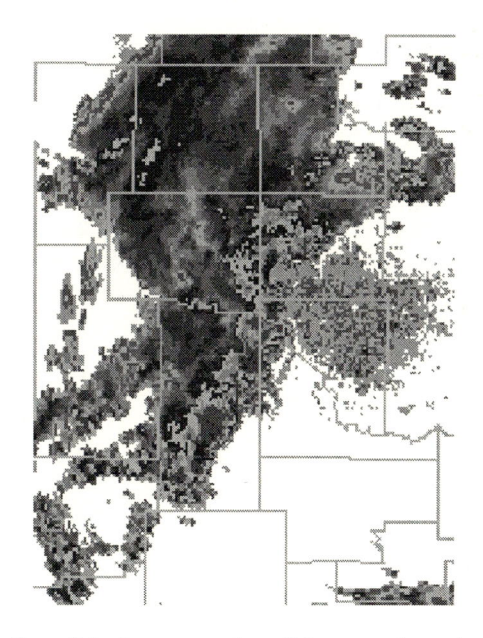

Figure 9-4. Appearance of squall line on weather radar in Oklahoma late in the evening of October 22, 2000. This squall line shows gaps and clusters, which shows that it also contains characteristics of multicell thunderstorms.

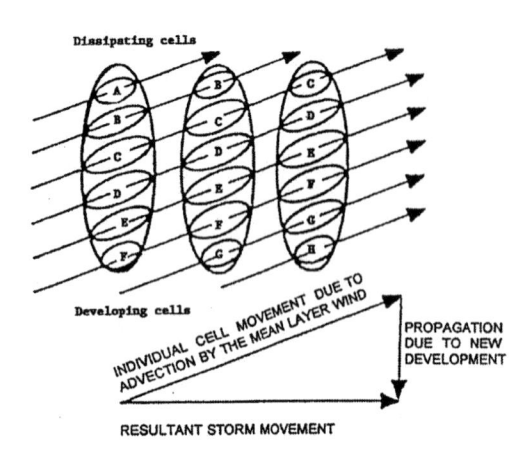

Figure 9-5. Concept of propagation of a multicell line. Even though the environmental winds may be flowing from southwest to northeast, new cell development ("backbuilding") occurs on the storm's flanking line to its south, with older cells to the north dying out. In effect, even though individual cells move toward the northeast, the storm as a whole moves eastward, to the right of the winds.

Figure 9-6. Three multicell lines in central Kansas on May 26, 2000, as seen by visible satellite imagery. The flanking lines are visible on the western and southern cells.

9.2.6. Bow echo. Sometimes, a cell within the squall line can "cheat" by accelerating ahead of its own neighbors. Such a bulge is called a bow echo, or in a stronger condition, an LEWP (line echo wave pattern). High winds and small hail can be expected ahead of the bow, with a slight chance of a weak, brief tornado (usually on the northern periphery).

9.2.7. Derecho. Occasionally, the entire squall line forms a bulge 50-200 miles wide, producing what is known as a derecho. These storms last several hours, and can produce a swath of 100 mph winds and hail in an area as large as a state. Derechos are most common in the midwest states (Illinois, Indiana, Iowa, etc) and are rare south of 32 degrees N latitude.

9.2.8. Split. Severe thunderstorm cells (particularly supercells) sometimes break in two, thought to be due to vertical updraft splitting by wake flow curl or precipitation loading. This is rare, but when it happens, the right split continues moving east and remains tornadic. The left split is weaker but can contain an anticyclonic tornado (very rarely though).

9.2.9. MCC. Stronger multicell clusters may develop in a large circular area, forming what is known as a mesoscale convective complex (MCC). These are often seen at nighttime during the summer over the central U.S., and may have smaller-scale characteristics of the unicell and squall-line thunderstorm. Agriculturally they are significant since they produce the majority of the warm-season precipitation in the Midwest. MCC's are quite dramatic on satellite, showing huge, cold anvils with strong divergence at 200 mb observable within the synoptic network, and may be readily observable on synoptic-scale surface charts. The storms normally sport hail and high winds, and may cause flooding since the multiple cells pelt a location continuously with rain. The MCC usually dies off by late morning.

9.2.10. MCS. The term mesoscale convective system (MCS) refers to any organized convective weather system, usually in the form of a squall line or a large, progressively-moving area of multicell storms. The system is large enough to disrupt synoptic-scale patterns, and the cirrus shield occupies a large area on satellite imagery.

9.3. SUPERCELL

The most dangerous type of thunderstorm in terms of the immediate threat to life and property is the supercell. It is defined as a convective thunderstorm with a mesocyclone or mesoanticyclone. The storm is basically a multicell which has exploited its environment to the fullest, shaping the surrounding air in a way which prolongs its own life.

The term "supercell" was invented in 1962 by Prof. Keith A. Browning, who to this day is still an active researcher at Reading University in Great Britain. The supercell storm develops in areas of high instability with some degree of shear (and sometimes in low instability areas with substantial shear — thus the value of the EHI Index, to be discussed later). The initial life cycle is similar to that of the multicell line, however the flanking line towers show a tendency to grow very close together, even merging into one another, and the updraft base is fed by a strong, persistent inflow. Often within an hour or two, the storm will further organize and may produce lowerings and tornadoes, and on radar may show hook-like appendages on the southern side.

9.3.1. Low-precipitation (LP) supercell. When very little precipitation falls out of a supercell storm, the result is called a low-precipitation (LP) supercell. Since the rain curtains and rain shafts remain small in area, vast portions of the supercell are exposed for viewing. This makes it excellent for photography. However, while the lack of rain shafts means great photography, their absence also takes away the processes that help spin up a tornado. So while these storms may produce only small tornadoes, they can still produce huge hail. Low-precipitation supercells are inefficient. This can be attributed to dry-air entrainment with high instability, in which the additional dry air makes the storm even drier. Also, strong shear with weak instability can produce an LP supercell, in which most precipitation forms the anvil rather than falls to the ground.

9.3.2. Classic (CL) supercell. This type of supercell is what's usually seen in textbooks and diagrams. It involves a moderate

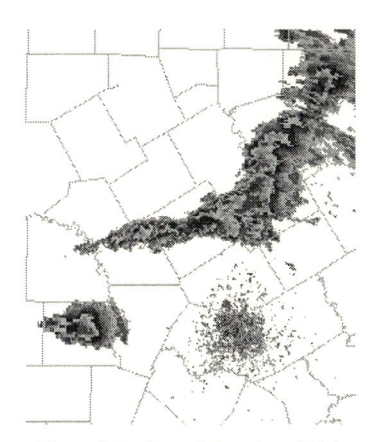

Figure 9-7. A weak derecho will take on an appearance like this, as seen on the afternoon of March 10, 2000 in central Texas. The isolated storm to the southwest also bears watching, as the stronger core is on the south side of the storm and it has the telltale triangular appearance of a supercell.

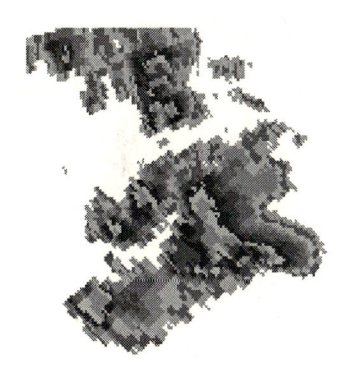

Figure 9-8. Supercell west of Zaragosa, Coahuila, Mexico on the evening of April 11, 2000. It is obscured slightly due to a weaker cell to the southwest "seeding" it. The combination of deep moisture and orographics sometimes makes northern Coahuila a powderkeg for supercells, particularly in March and April.

amount of precipitation that falls out of the storm — just enough to help tornadoes spin up. These rain curtains get wrapped into the mesocyclone circulation, producing the hook echoes observed on radar. However, these rain curtains are eventually the tornado's demise. They get wrapped into the circulation and occlude the tornado. This type of storm results from high instability and large storm-relative shear.

9.3.3. High-precipitation (HP) supercell.

The high-precipitation (HP) supercell can be one of the most violent types of storms on Earth. Copious amounts of precipitation fall out of the storm. Weak shear is often what causes the storm to be a HP supercell, because it allows the precipitation to fall back into or just outside the updraft, resulting in rain-wrapped tornadoes. Where there is somewhat stronger shear, precipitation falls around the updraft, but is wrapped into the mesocyclone, forming what is known as a "bear's cage" (the tornado is hidden inside the cage of rain).

9.4. WIND PROFILES

The character of a thunderstorm is largely determined by its storm-relative wind profile. To evaluate storm-relative winds, we need the hodograph, a tool which was described in Chapter 3, Analysis.

Miniature supercell?

Figure 9-9. Structure of typical classic supercell. The wall cloud and potential tornado are located at the circle with the "X" in it. The small-scale fronts are indicated by the frontal symbology.

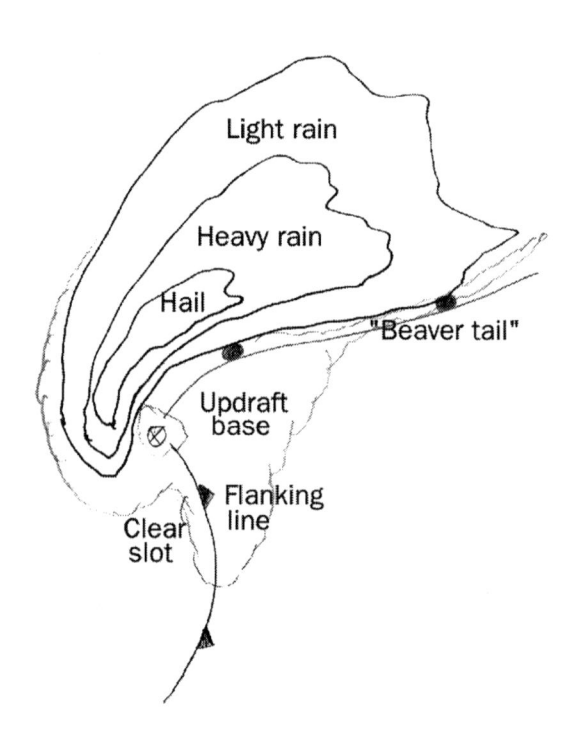

The basis of the hodograph is quite simple: upper-level winds are always measured relative to the ground. When there is a 30 mph wind blowing from west to east across the plains, we assume it is blowing over barns at 30 mph, over treetops at 30 mph, and over lakes at 30 mph. However if there is a storm moving eastward at 40 mph, it would see this wind as moving from east to west (the opposite direction!) at 10 mph. Therefore one of the most important goals of working with a hodograph is to evaluate storm motion. Once storm motions have been established, the amount of shear, inflow, and ventilation can be determined. These help indicate what type of storm will occur.

It is important to not just use the most convenient hodograph that's available. The hodograph must be modified to provide the best possible estimate of winds near the storm threat area at the time of initation. Numerical model output, radar VAD/VWP winds, and profiler winds can all contribute greatly to this task.

9.4.1. Storm motion.
In some cases, storm movement can be determined simply by studying radar animations. However the atmosphere rarely provides this luxury, requiring the forecaster to anticipate storm character before any activity even develops.

In general, the movement of a thunderstorm tends to represent an average of the winds at all levels through the troposphere. However it has been found that the lower levels of this depth are most important. Therefore the standard method of determining storm motion is to take an average of the 0-6 km winds. This can be done on the hodograph by visually averaging all the points between 0 to 6 km, ensuring that an even distribution of heights from 0 to 6 km are considered.

For typical storms, this techniques works quite well. Unfortunately storms often deviate from this movement and complicate the profile. They need to be planned for, and the deviant movement must be plotted on the hodograph.

For at least a decade, a rule of thumb has said that when normal storm movement is less than 30 knots, the deviant storm moves 20 degrees to the right and 85% the speed of the regular storm motion (thus the 20R85 rule). If normal storm movement is greater than 30 knots, the deviant storm moves 30 degrees to the right and 75% of the speed of the regular storm motion (thus the 30R75 rule). Unfortunately this rule of thumb sometimes performs poorly since it performs computations relative to ground speed.

A new method, called the ID method, calls for deviant motion to be evaluated not relative to the ground but to the 0-6 km shear vector of the storm. The forecaster simply draws the normal storm motion dot, draws a shear vector connecting the average 0-0.5 km and average 5.5-6 km wind vectors, and draws a

Figure 9-__. **Capt. Robert C. Miller** (1920-1998) and **Maj. Ernest J. Fawbush** (1915-1982). These two U.S. Air Force forecasters were the first to issue any type of successful advance tornado forecast that was officially sanctioned. The forecast, which provided just over three hours of advance notice, was issued at Tinker AFB, Oklahoma on March 25, 1948 under the authority of General Fred S. Borum. In 1951 Miller and Fawbush were instrumental in establishing a centralized severe weather forecasting facility, followed the next year by a similar facility in the National Weather Service which issued the first public watch box. In 1972 Miller published an impressive set of guidelines for forecasting severe weather, Notes on Analysis and Severe Storm Forecasting Procedures of the Air Force Global Weather Central, during his long tenure there as both an officer and a civilian researcher.

line perpendicular to this line which intersects the normal storm motion dot. The deviant movement dot is plotted along this second line at a vector length of 7.5 m/s (15 kt) from the storm motion dot on each side of it. The deviant movement dot on the right side of the 0-6 km vector (facing the 6 km dot from the 0 km dot) is the right movement vector, and the other one is the left movement vector.

Still, storm movement is not an easy task and can be complicated by frontal boundaries, cold pools, topography, and unusual storm structures. Errors can also occur from poorly constructed hodographs using nonrepresentative winds, or unusual layers of winds in the atmosphere.

From here on, "storm motion" will refer to the movement of a storm along either the normal or the deviant movement vector. It's up to the forecaster to assess these possibilities and what they will contribute to storm development.

9.4.2. Storm-relative inflow. By comparing the length of the vector between the storm motion vector and the lowest 2 km of the atmosphere, this yields the storm-relative inflow. The greater the inflow, the greater the feed of moisture into the storm. Storms with weak storm-relative inflow tend to be weak.

9.4.3. Storm-relative helicity. The geometric area swept out between the storm motion vector and the lowest 2 km of the hodograph defines the storm-relative helicity. This is one of the key ingredients tied to rotating storms and tornado development. A computer can quantify these values, but a trained eyeball estimate can suffice quite well.

One configuration that can cause a larger area to be swept out is a curved hodograph trace, particularly one which is curved in the lowest levels of the atmosphere. This allows the hodograph trace to wrap more fully around the storm motion point. Backing winds at the surface and in the lowest level of the atmosphere can allow a relatively straight hodograph to take on curvature that increases storm-relative helicity values.

A straight-line hodograph tends to favor splitting storms, where a severe storm splits and one portion moves along the right deviant motion vector and the other moves along the left.

9.4.4. Storm ventilation. By comparing the length of the vector between the storm motion vector and the upper part of the atmosphere, this indicates the speed of the anvil cloud and precipitation relative to the storm (i.e. the ventilation of the storm). Large storm-relative ventilation values are essential for a long-lived updraft.

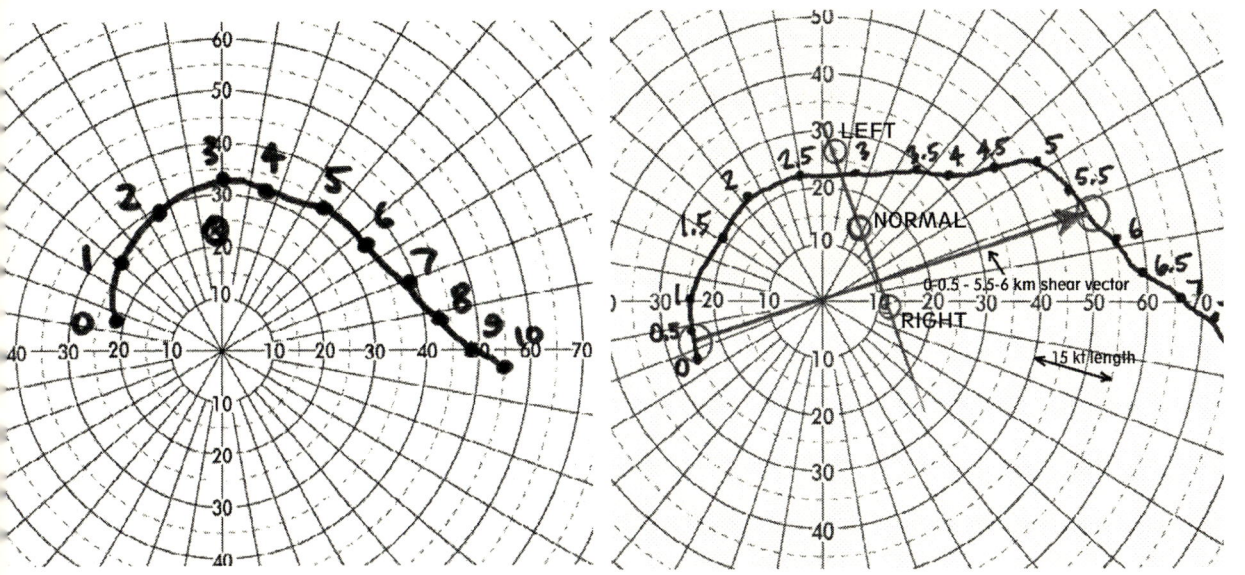

Figure 9-10a. Calculation of normal storm motion. The storm motion corresponds to the geometric center of the profile trace in the lowest 6 km of the atmosphere. This can be calculated mathematically, but it's faster to do it mentally by visualizing a spot that could "gravitationally" balance each of the points from 0 to 6 km. In terms of ground motion, this particular storm motion example illustrates cell motion moving north at 25 knots (originating from 180 deg and moving towards 360 deg).

Figure 9-10b. Calculation of deviant storm motion, ID method. A vector is drawn connecting the average 0-0.5 km and 5.5-6 km wind. Then a line is drawn perpendicular to this vector that intersects the normal storm motion. A right and left deviant movement vector is then plotted 15 kt on either side. It can be seen that right-moving storms will move eastward at 10 kt while left-movers will track northward at 25 kt.

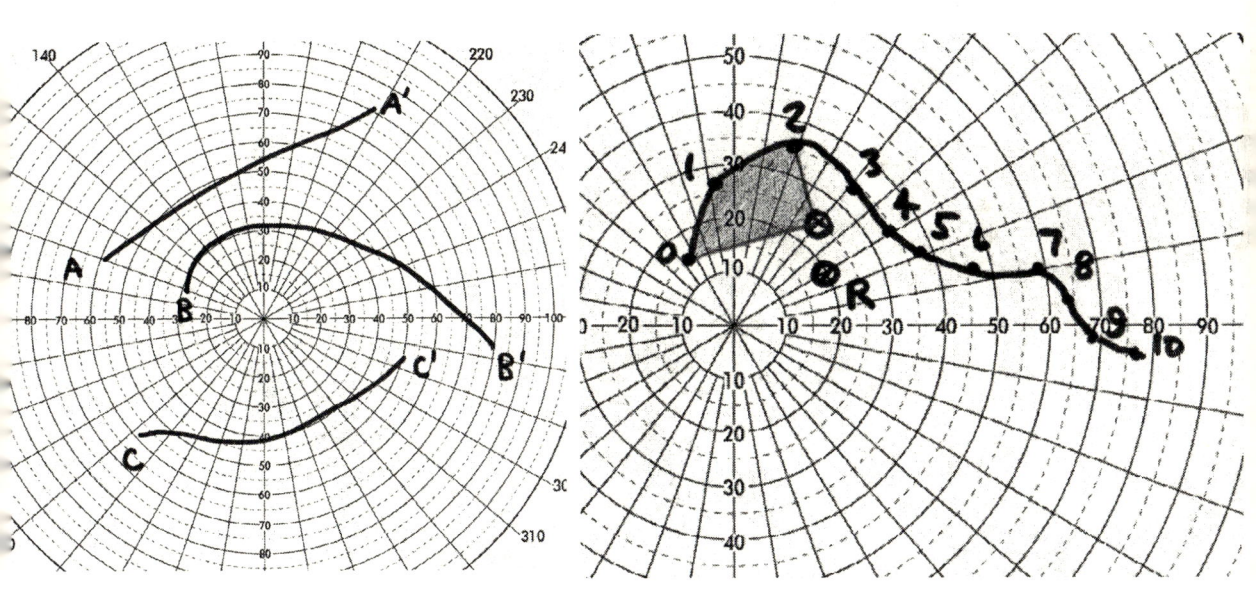

Figure 9-10c. Depictions of straight and curved hodographs. Trace A-A' is a straight hodograph that favors splitting storms and squall lines. Trace B-B' is a clockwise-curved hodograph that favors right deviant motion and contains positive storm-relative helicity. Trace C-C' is a counterclockwise-curved hodograph, rare on storm days, that favors left deviant motion and contains negative storm-relative helicity.

Figure 9-10d. Storm-relative helicity is the geometric area swept out between the storm motion vector and the hodograph trace between 0 and 2 km. The larger the area, the larger the storm-relative helicity. Note that if a storm begins moving along the right-movement deviant vector (labelled "R") the area swept out will be much greater.

9.5. INSTABILITY PROFILES

Since the 1950s, forecasters have attempted to quantify instability and other parameters to provide a consistent methodology for anticipating severe weather. Many of these techniques are highly useful, however they have various drawbacks. One is that a predictor which uses two or three levels may completely overlook a significant feature, such as an inversion or a layer of moisture. Furthermore, the sounding must be modified to account for conditions at convection time before applying the methods; the CAPE for a 1200 UTC sounding is often unrepresentative of storms that develop at 2100 UTC (a common pitfall when using Internet products). Overall, common sense must be your guide when using these methods.

9.5.1. Convective available potential energy (CAPE), B+. The CAPE value is by far one of the most useful predictors for thunderstorms. It equals the vertically integrated positive buoyancy of an adiabatically rising parcel throughout the entire troposphere. In fact, updraft speed is closely tied to CAPE. Any

Figure 9-11. The updraft of a thunderstorm is not directly visible to radar but represents the floodgate for inflow and moisture streaming into the storm and rising. It is the part of the storm that harbors the mesocyclone and produces tornadic activity.

positive CAPE value indicates that rising motion will occur, providing any lower inversions can be overcome. A values of up to 1000 J/kg indicates weak convection, 1000 to 2500 J/kg indicates moderate convection, and above 2500 J/kg indicates strong convection.

9.5.2. Convective Inhibition (CIN), CINH, B-. This value represents the negative energy area on a SKEW-T below the level of free convection, and is equivalent to negative CAPE. If natural processes fail to destabilize these negative areas, energy from forced lift will be required to move the negatively buoyant parcels upward to their LFC. Like CAPE, CIN is measured in J/kg. The significance of CIN depends on the buoyancy in layers below and the amount of available forced lift that can occur.

9.5.3. BRN Shear (bulk shear). This is simply the vector difference between the 0-6 km wind and the near-surface winds. Its chief use is to determine whether storms will be long-lived (if the downdraft will be separated from the updraft).

9.5.4. Bulk Richardson Number (BRN). This is defined by CAPE divided by the BRN Shear. This attempts to combine instability and bulk shear in order to separate storm types into various categories. BRN's less than 45 tend to support supercell structures (values below 10 tend to shear out any developing convection), while multicellular convection favors BRN's above 45. BRN by itself is a poor predictor of storm rotation, since it only addresses bulk shear (which determines a storm's longevity) and not low-level helicity (which allows for rotating storms). BRN yields the following results given a certain amount of instability and bulk shear:

Instability	Bulk Shear	BRN result
Low	High	Low
Low	Low	Medium
High	High	Medium
High	Low	High

9.5.5. Storm-relative helicity (SRH). This is a summation of streamwise vorticity through the storm inflow layer, and is dependent on figuring an appropriate value of storm motion (typically a value 30 degrees to the right of the 0-6 km wind with 75% of its magnitude, "30R75"). The SRH value is generally determined using computer diagnostics.

9.5.6. Energy-helicity index (EHI). This equals the product of positive SRH and CAPE, divided by 160,000. It is an attempt to combine the effects of CAPE in producing strong updrafts and

---

HUMOR BREAK: Thumb tab tornadoes
I've always loved the Audubon Society field guides for their exceptional information and superb photographs, but while reading "How To Use This Guide" in *Field Guide to North American Weather* (1991), it seemed they were trying too hard to push the "thumb tab" concept: "You are visiting friends in Kansas... you scan the sky and notice that a small funnel has formed... 1. Turn to the Thumb Tab Guide. There you will find a vortex-shaped silhouette, standing for the group Tornadoes and Other Whirls. The symbol refers you to the color plates 207-230. 2. You look at the color plates and quickly surmise that the funnel shape in the sky overhead may be the beginning stage of tornado. The captions refer you to the text on pages 511-522. 3. Reading the text, you become convinced that there is a tornado forming. You and your friends immediately seek the storm shelter and wait for the threat to pass."

storm-relative helicity in providing rotating updrafts. EHI values above 1 indicate strong tornadoes, while violent tornadoes favor EHI values of 5 or greater.

9.5.7. K-Index (KI). Equals $K = (T_{850} - T_{500}) + Td_{850} - Tdd_{850}$, where Td is dewpoint and Tdd is dewpoint depression. This represents thunderstorm potential as a function of vertical temperature lapse rate, low-level moisture content, and depth of the moist layer. It has been found that KI is good for forecasting heavy rain but not for determining whether storms will be severe. Values between 31 and 35 tend to favor scattered storm development, with higher values indicating numerous thunderstorms.

9.5.8. Lifted Index (LI). This is simply the Celsius degree difference in temperature between a representative (mixed) low-level parcel lifted to 500 mb and the environmental 500 mb air. A parcel that is warmer than its environment has a negative LI. Values of -5 or less are associated with severe weather.

9.5.9. Showalter Index (SI). This is the Celsius degree difference in temperature between an 850 mb parcel and the environmental 500 mb air. Since moisture depth is not taken into account, it is not a preferred method for figuring storm potential (the Lifted Index should be used). Values of -4 or less are associated with severe weather.

9.5.10. Severe weather threat index (SWEAT). This combines low level moisture, instability, jet speeds, and warm advection. Values above 300 indicate a potential for severe storm development, and above 400 indicate tornadic storms. The value is computed as:
$$\text{SWEAT} = 12(Td_{850}) + 20(TT-49) + 2(V_{850}) + V_{500} + 125(\sin(D_{500} - D_{850}) + 0.2)$$
where D is the wind direction and V is the wind speed

9.5.11. Total totals index (TT). Combines lapse rate and low level moisture to estimate the potential for severe convection. When the TT exceeds 50, a few severe storms are indicated. Values above 52 favor scattered to numerous thunderstorms, and values above 56 support numerous thunderstorms with scattered tornadoes.
$$TT = (T_{850} - T_{500}) + (Td_{850} - T_{500})$$

9.5.12. Vertical totals index (VT). Determines the lapse rate.
$$VT = T_{850} - T_{500}$$

9.5.13. Cross totals index (CT). Relates low-level moisture to mid-level temperatures.
$$CT = Td_{850} - T_{500}$$

9.5.14. Wet-bulb zero (WBZ). This indicator, generally used for hail prediction, equals the lowest height *above ground level* where the wet-bulb temperature is below 0 deg C. Hail does not generally melt when the wet-bulb temperature is below zero, so when this level is closer to the ground, hail is less likely to melt as it falls, with the result being larger hailstones. WBZ heights of less than 10500 ft AGL correlate well with large hail at the surface when storms develop in an air mass primed for strong convection.

REVIEW QUESTIONS

1. What situation can cause thunderstorm downdrafts to become very dense, producing strong and damaging outflow?

2. How does the updraft orient itself in a squall line storm?

3. Why are tornadoes rare in most squall lines?

4. Is damaging weather more likely in a squall line if it organizes and solidifies or if it breaks up into isolated cells?

5. Where is the most dangerous weather found in most types of storms, relative to the updraft and downdraft?

Figure X-XX. The Storm Prediction Center in Norman, Oklahoma is responsible for severe thunderstorm and tornado watches for the United States. It is this desk from where they are issued.

6. While monitoring the weather radar, you notice a 10-mile long section of a squall line moving slightly ahead of the line, developing a rounded appearance. What is this feature and what does it suggest?

7. What type of process usually leads to two adjacent supercells, each a mirror image of each other with a rotating updraft?

8. Identify one of the most significant dangers of LP (low-precipitation) supercells.

9. Why are HP (high-precipitation) supercells more dangerous and difficult to monitor than typical supercells?

10. Which prediction indicator gives the most accurate reflection of instability throughout the entire troposphere?

Convective snow showers, high winds, and 10°F (-12°C) temperatures rake the Minnesota prairie. *(Tim Vasquez)*

10 WINTER FORECASTING

"It's snowing still," said Eeyore gloomily.
"So it is."
"And freezing."
"Is it?"
"Yes," said Eeyore. "However," he said, brightening up a little, "We haven't had an earthquake lately."

ALAN ALEXANDER MILNE
The House at Pooh Corner, 1928

Winter weather brings a forecast challenge almost equal in complexity to that of severe thunderstorms. The stakes are high: rather than a few localities being raked by high winds or a tornado, an entire state may be buried in inches of ice or snow. Interestingly some of the techniques for forecasting winter weather have a close kinship with that of severe weather: close examination of soundings, careful diagnosis of upper-level lift, and hour-by-hour analysis of surface charts.

Millions of crystals
Renowned cloud physics expert Vincent J. Schaefer estimated that it takes more than half a million ice crystals to cover a one-square-foot area with snow ten inches deep.

10.1. PRECIPITATION TYPE

Skill for forecasting winter weather depends, of course, on a firm knowledge of the types of winter precipitation that occur and how they develop. Therefore the chapter will begin with such a summarization. Note that a "cold" layer used in a winter forecasting context refers to subfreezing temperatures, and a warm layer refers to temperatures above freezing.

10.1.1. **Rain** is a liquid form of precipitation. Its presence implies that there are significant layers of warm air in the lower troposphere.

10.1.2. **Snow** is a solid form of precipitation. It is composed of ice crystals and requires a cold atmosphere. The snow cannot pass through a warm layer any deeper than 600 ft, otherwise partial melting will occur. Precipitation that has melted cannot turn back into snow.

10.1.3. **Ice pellets (sleet)** occurs when liquid precipitation freezes <u>before</u> striking the ground.

10.1.4. **Freezing rain** occurs when liquid precipitation freezes <u>after</u> striking the ground. A warm ground temperature may inhibit freezing even though air temperatures are cold.

10.2. HORIZONTAL ANALYSIS

The forecaster must first evaluate the atmosphere for precipitation potential, regardless of type. This is done by looking for sources of upward vertical motion, as described in Chapter 6, as well as for convective sources due to instability. Favored areas for upward motion relative to a classic extratropical cyclone are along and north of the warm front and in proximity to the low pressure center.

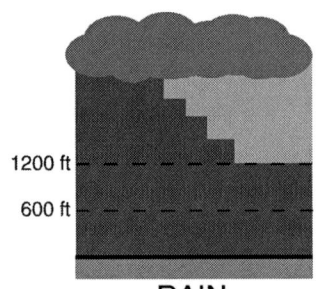

144 • WEATHER FORECASTING HANDBOOK

Figure 10-1a. Rain. Warm (above-freezing) temperatures are the dark shade on the left, and cold (below-freezing) temperatures are the light shade on the right.

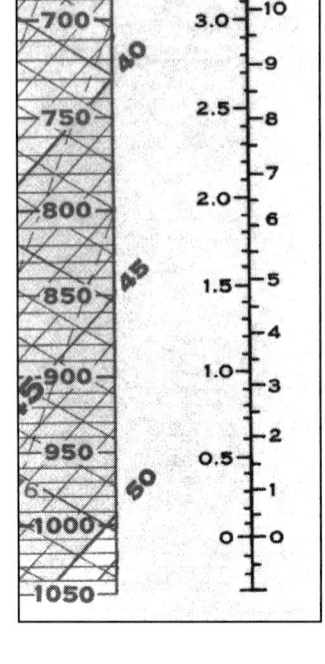

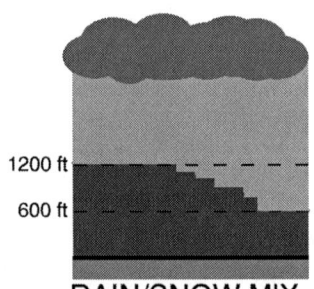

Figure 10-1b. Rain/snow mixes are favored in cold atmospheres where a 600 - 1200 ft deep warm layer is at the surface.

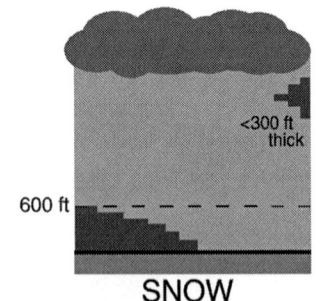

Figure 10-1c. Snow occurs in cold atmospheres where any warm layers are negligable.

Figure 10-2. Altitude scale on Skew-T sounding. Winter forecasting requires the forecaster to make delicate estimations of layer depth. The altitude scale is highly useful for converting sounding traces and millibar heights (left) to meter/feet heights (right). Unfortunately many Internet-based Skew-T's omit the altitude scale and use undersized graphics that make the lower part of the trace difficult to read, which seriously hampers the ability of a forecaster to apply winter forecasting techniques. If necessary, the forecaster must use a software-based sounding visualization program or plot the sounding on paper.

If the charts show that warming or cooling is expected in a particular area, the forecaster must file this information away for use when performing the vertical evaluation.

Once the precipitation potential has been established, then the precipitation type can be considered. If no precipitation is forecast, there is no need to go any further.

10.3. VERTICAL ANALYSIS

The forecaster must have a thorough knowledge of the vertical thermal structure of the atmosphere that's being forecast. For this, it is important to look at all available soundings and to be able to modify the soundings with surface data and interpolation methods. Finally the forecaster needs to be proficient at estimating altitudes on the sounding chart, which allows accurate estimations of vertical thicknesses.

10.3.1. Rules of thumb. The situations that follow are based on the following rules of thumb:

■ Solid that passes through a warm layer depth of at least 600 ft will melt partially.

- Solid that passes through a warm layer depth of 1200 ft will completely melt.
- Liquid that passes through a cold layer depth of 800 ft will freeze.
- Liquid that refreezes will always form a frozen droplet. Snow is not formed in this manner.

10.3.2. Cold atmosphere. When the entire column of air above a station is below freezing, snow will be the result. The character of the snow is explained in section 10.5 (Snow Forecasting).

10.3.3. Warm atmosphere. When the entire atmosphere up to about 2000 feet is above freezing, the forecast is quite simple: forecast rain! Any precipitation that falls through the 2000 ft thick warm layer will melt as it falls. However the forecaster must be alert to the presence of dry layers which may chill layers of the atmosphere and alter the temperature profile.

10.3.4. Warm aloft; cold at the surface. This situation occurs when warm air overruns subfreezing air in the lower levels, and is most frequent in the region ahead of a warm front. It is the most complicated situation for predicting winter precipitation.

10.3.5. Cold aloft; warm at the surface. This type of situation usually occurs when a strong, cold upper-level system (such as a cutoff low or strong short wave) works across an airmass with mild temperatures. This causes steep lapse rates that allows air to cool rapidly with height.

The precipitation will almost always form as snow, so the question is will it reach the surface before melting into rain? Studies have found that if the warm layer is more than 1200 ft deep, it will most likely rain. If it is around 900 ft deep, the chance of it snowing instead of raining is 50%, if it is 700 ft deep, the odds increase to 70%, and at 300 ft deep, the odds are 90%. The freezing level often lowers 500 to 1000 feet during the first 1-2 hours after the precipitation begins falling, and may rise again to its original level 3 hours after the layer becomes saturated. Also consider the situation described in the next section, about rich upper-level moisture falling into very dry low-level conditions.

10.4. DIABATIC CHANGES

10.4.1. Evaporation. Changes in the column of air over a station due to evaporation can be massive and can overwhelm all other processes. This is especially true when northerly winds are

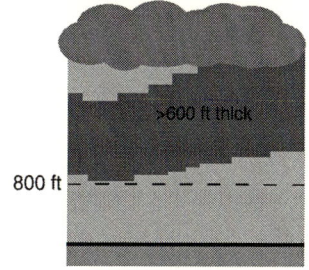

ICE PELLETS

Figure 10-1d. Ice pellets occur in a warm atmosphere when a layer of cold air greater than 800 ft deep is at the surface.

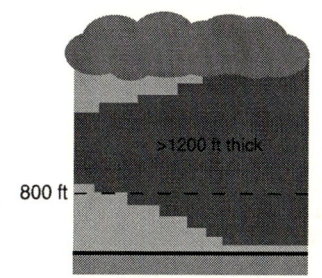

FREEZING RAIN

Figure 10-1e. Freezing rain occurs in a warm atmosphere where a layer of cold air shallower than 800 ft is at the surface.

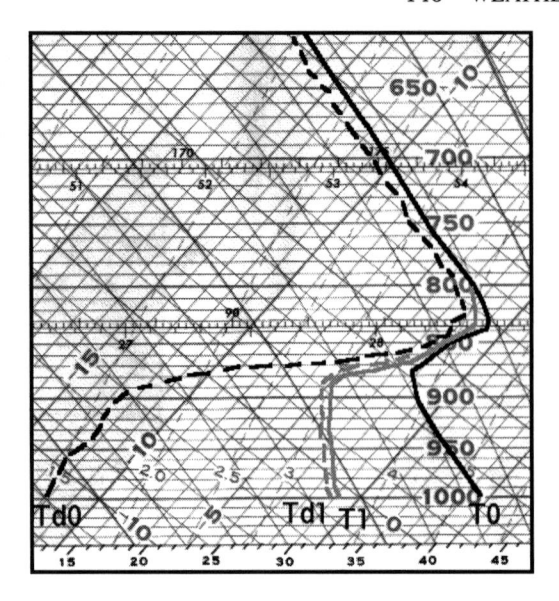

Figure 10-3. The effect of evaporative cooling on a sounding. The initial sounding is represented by the dark T0 and Td0 lines. Note the surface temperature of 4 deg C (39 deg F), the layer of warm air is at least 2000 ft deep (guaranteeing rain), and how the dry air occupies a layer between the surface and 875 mb. As precipitation falls into this dry air, evaporative cooling occurs. Some rain initially reaches the surface. Over the following hour or two, depending on the precipitation rate, the column cools to its wet bulb temperature. When "wet bulbing" is complete, the sounding profile is T1 and Td1. The entire column is subfreezing, snow is the precipitation type, and the surface temperature has cooled to -1 deg C (30 deg F).

injecting cold, dry air at the low levels, with precipitation layers developing aloft due to upper-level or isentropic lift.

When snow begins falling, the dry air in the low levels will cause the snow to evaporate as it falls, producing virga. This evaporation process chills this dry air, cooling it to its wet bulb temperature (which is usually about halfway between the air temperature and the dewpoint temperature). Ultimately the entire column of air can cool to a subfreezing temperature, changing what looked like 38 deg F with rain to 29 deg F with snow! This process has been responsible for massive forecast busts for unwary forecasters, including a record breaking snowfall in Abilene, Texas in January 1995.

It has been shown that only 0.38 inches of liquid equivalent precipitation can contribute 8 deg F of cooling towards an air parcel's wet-bulb temperature!

10.4.2. Melting. When large amounts of snow or ice fall through an elevated warm layer and melt, this process absorbs heat from the warm layer. As a result, the depth and temperature of the warm layer diminishes. In actuality, temperature advection and even evaporation tends to overwhelm this diabatic process but in marginal situations with weak advection it can become a substantial process.

If warm-layer erosion due to melting is actually taking place, the net effect with time is that the surface precipitation changes from rain or freezing rain to sleet and mixed precipitation, and from that into snow.

Figure 10-4. Dendritic hexagon snowflake as seen under an electron microscope. This flake was likely produced by accretion as it fell through the cloud. (USDA/BARC)

10.5. SNOW CHARACTERISTICS

10.5.1. Crystal shapes. Ice crystals can form several distinct shapes which are associated with various ranges of formation temperature.
- Thin plates: 0 to -4 deg C.
- Columns, prisms, needles: -4 to -10 deg C.
- Thick plates: -10 to -12 deg C.
- Dendrites: -12 to -16 deg C.
- Sector plates: -16 to -22 deg C.
- Hollow columns and sheaths: below -22 deg C.

10.5.2. Heterogeneous nucleation (liquid to solid). This is the basic step in ice crystal formation within a snow cloud. Droplets of supercooled liquid water, which commonly exist in saturated layers as cold as -20 deg C, freeze directly on aerosols and dust. The process is most efficient at temperatures colder than -10 deg C. The particle that this produces is quite small and requires additional stages of growth before it can be considered significant for any forecasting purposes.

10.5.3. Deposition (gas to solid). This stage of growth allows water vapor to condense onto ice crystal nuclei. This process works best at -20 to -10 deg C (particularly in the -15 to -12 deg C range). Crystals that develop are still too small to be of forecast significance, generally less than 1 mm.

10.5.4. Ice multiplication (riming or accretion). When ice crystals drift into warmer temperature layers (-10 to 0 deg C), supercooled liquid droplets freeze directly upon these crystals. This produces large, fragile crystals that splinter into pieces, providing another set of condensation nuclei. This results in a

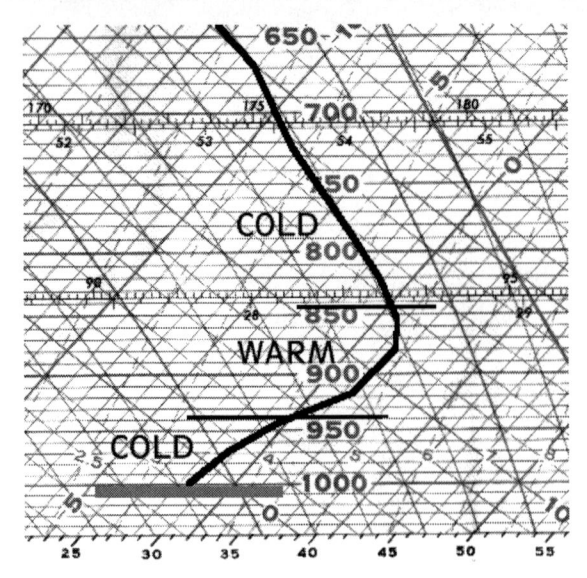

Figure 10-5. Ice pellets are associated with a sounding like this. The surface cold layer extends from 1000 to 960 mb, suggesting a depth of 2000 ft. This provides a vast layer in which liquid precipitation can refreeze.

chain reaction of ice crystal growth. This is the mechanism by which most snowflakes develop. Excessive riming can cause graupel and sleet. When all temperatures in a column above a station have fallen below -10 deg C, this indicates that ice multiplication will cease and measurable snowfall amounts will taper off.

10.5.5. Aggregation. When only one type of ice particle (dendrite, column, etc) is in a cloud, the particles tend to fall at similar speeds. However when there are multiple types of ice particles, fall speeds are varied and the chance of ice crystal collision is greatly enhanced. This results in growth of the ice crystal, turning it from a ice crystal into a snowflake and making it vastly larger by the time it reaches the ground. The best way to produce multiple types of ice particles (and to produce snowflakes) is to have a broad range of temperatures within the saturated layer (preferably the full range from 0 to -20 deg C). Furthermore, it has been found that an isothermal layer at about -5 to 0 deg C extending through 1,000 ft or more can further boost aggregation of ice crystals, resulting in very large snowflakes. This is because snowflakes tend to be "sticky" at these warmer temperatures.

REVIEW QUESTIONS

1. What special process of ice crystal or snowflake formation is most likely responsible for heavy snowfall events?

2. Rain falls into a layer of cold air 500 feet deep, then strikes the ground. What type of precipitation is occurring at the ground?

3. Snow falls into a layer of warm air 1500 feet deep, then into a layer of cold air 1000 feet deep, where it then strikes the ground. What type of precipitation occurring at the ground?

4. Rain falls into a layer of cold air 500 feet deep, then into a layer of warm air 500 feet deep. What type of precipitation is occurring at the ground?

5. Freezing rain has been occurring all morning. The local news channel's towercam, 1000 feet up on the transmission tower, is beginning to ice up. What changes in precipitation type can be expected?

6. It is 30°F (0°C) and the sounding shows no inversions. What is the most likely precipitation type?

7. A warm front is south of Kansas City, which is currently reporting ice pellets (sleet). As the warm front approaches, what precipitation type transitions can be expected?

8. Is the temperature profile in a cold-core barotropic low conducive to large snowflake formation?

9. Why is heavy snow rare when the column of air above a station is below -10 degrees Celsius?

10. Ice pellets have been occurring all day. A television station's engineers need to climb a 1000 ft television tower to perform some maintenance. Will the engineers most likely encounter snow or freezing rain? Is it safe for them to perform the task?

Hurricane Andrew near Louisiana on August 25, 1992. *(NASA)*

11 TROPICAL WEATHER

I looked forward to the coming of the monsoon and I became a watcher of the skies, waiting to spot the heralds that preceded the attack. A few showers came. Oh, that was nothing, I was told; the monsoon has yet to come. Heavier rains followed, but I ignored them and waited for some extraordinary happening. While I waited I learnt from various people that the monsoon had definitely come and established itself. Where was the pomp and circumstance and the glory of the attack, and the combat between cloud and land, and the surging and lashing sea?

JAWAHARLAL NEHRU
The Monsoon Comes To Bombay, 1939

In the tropics, weather is dominated by convective activity and the lack of baroclinicity. Occasionally a cold front will make it far south and produce genuine frontal weather, but such occurrences are somewhat rare. Furthermore, since pressure gradients are quite weak, forecasters must resort to streamline analysis to keep track of weather patterns. Streamlines are drawn parallel to the wind flow and can uncover troughs, ridges, cyclones, anticyclones, outflow areas, and other disturbances.

11.1. EQUATORIAL TROUGH (ITCZ)

There is always a belt of low pressure lying between the subtropical ridges of the north and south hemispheres, called the equatorial trough or the intertropical convergence zone (ITCZ). It more or less follows the day-to-day latitude of the sun, shifting north in the summer and south in the winter. The region is the focus of convergence between the subtropical ridges, enhancing upward motion and favoring the development of clouds and rain. These frequently appear on satellite imagery as an east-west band of clouds.

11.1.1. Trade wind trough. This type of equatorial trough is associated with convergence between the easterly trade winds of the north and south hemisphere. It is most common over open ocean areas away from the influence of continental heating. Occasionally a wave may develop along this trough, moving westerly at 10 to 15 kts; it is called an equatorial wave. Equatorial waves bring enhanced convection along the equatorial trough.

11.1.2. Monsoonal trough. This type of equatorial trough occurs when its position is affected by strong insolation of land surfaces. Warm-core barotropic lows develop in the region of strongest heating, focusing convergence. As deeper and deeper moisture advects poleward into these lows over a period of weeks, larger areas benefit from convective rains. One of the most common locations for monsoonal troughs is India.

11.1.3. Equatorial wave. This is a migratory wave-like disturbance in the equatorial trough. It usually moves from east to west at about 10 to 15 kts. Convection is enhanced where it occurs, and the nearby equatorial trough is often referred to as an "active ITCZ".

Author's tropical memories
— August 3, 1979 . . . flying from Guam to the Philippines at 32,000 feet, the weather was unlike anything I've ever seen in the United States. The sky was simply massive. Phenomenal cumulus towers reached far past our flight level, easily topping out at 50,000 feet. For over a thousand miles our DC-8 had no choice but to weave its way through and around them. Peering out the window at hundreds of clouds twice the size of Mount Everest, I felt very small. Little did I know that ten days later the deepest tropical cyclone ever recorded would tear through this area.
— August 1994 . . . in the city of Mombasa, Kenya along the Indian Ocean coast, myself and another forecaster were waiting for a bus one sunny afternoon. As we waited we saw a fuzzy line of low cumulus moving in from the sea. Even though the locals were heading indoors, we thought nothing of it. A few minutes later the cloud passed over and a torrential rain started. It was over in two minutes but we were thoroughly soaked.

TIM VASQUEZ

11.2. SUBTROPICAL RIDGE

This is the descending part of the Hadley cell (see the section on global circulation in the first chapter). It is produced when air in the equatorial trough rises, moves poleward, converges due to the Coriolis force, and sinks. The subtropical ridge's average position is 30 degrees latitude, and it contains vast amounts of subsident air. Therefore it is frequently sunny and warm under this ridge. High pressure areas often form within this ridge, which are warm-core barotropic systems.

11.3. TRADE WINDS

Air flowing equatorward from the subtropical ridge forms the trade winds, which dominates tropical latitudes. Since the air deflects to the right because of the Coriolis force (left in the southern Hemisphere), the result is a large band of easterlies: northeasterlies north of the Equator and southeasterlies south of the Equator.

11.3.1. **Air masses**. The deepest tropical air, usually within 15 deg of the Equator, is referred to as the "deep easterlies". Wind is consistently easterly throughout most of the atmosphere. Further north and south are the "shallow easterlies", which are only 10 to 20 thousand feet deep. The air above the shallow easterlies usually shows westerly flow, due to the effects of the mid-latitude (Ferrell) circulations.

11.3.2. **Trade wind inversion**. Because of the origin of this air from poleward latitudes, a cool-below-warm thermal setup occurs within the trade winds, forming what is known as the trade wind inversion. The vertical temperature difference can be as much as 10 deg C, and vertical cloud development is largely suppressed. Areas within the trade winds with strong equatorward flow, particularly on the east side of subtropical high pressure areas, see the strongest trade wind inversions.

11.4. SUBTROPICAL STATIONARY FRONTS

In eastern Asia, where it is rare for strong upper-level disturbances to emerge from mainland China, it is quite common for a boundary between the cool polar high in eastern Russia and the Pacific subtropical high to stagnate in one general location for weeks at a time. This east-west stationary front,

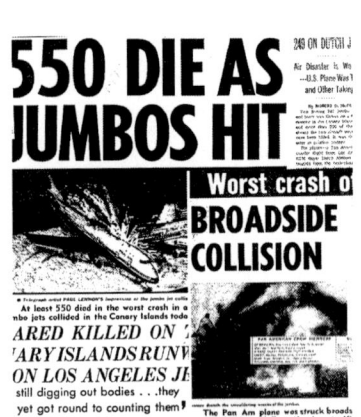

Figure 11-1. The collision of a KLM 747 with a Pan Am 747 on March 27, 1977 on a fog-shrouded runway in the Canary Islands still ranks as the worst air disaster in recorded history. At a latitude of 28 degrees, the Canary Islands are frequently under the northern fringes of the trade wind inversion. This stable air mass is a perfect breeding ground for persistent clouds and fog.

The constancy of the trades
In the trade [wind belt] the constancy attains 80 percent. In no other regime on earth do the winds blow so steadily. Life has adjusted to this uniform wind stream in numerous ways. On many islands the towns lie on the leeward side, which affords better protection to shipping from wind and ocean. The author recalls from his days in Puerto Rico that there was always a great deal of excitement in the office during one of the rare "interruptions of the trade".

HERBERT RIEHL
"Tropical Meteorology", 1954

sometimes incorrectly called the "monsoon", is known as the Changma by Korean forecasters. It tends to linger around Taiwan and Okinawa during the spring months, shifting north to Japan in late spring and to Korea in early summer as the subtropical high gains strength. Precipitation is frontal in nature and is frequently convective, with much of the upward motion originating from isentropic lift. Subtropical stationary fronts may occasionally stagnate in the southeast U.S., but normally the result is rapid cyclogenesis after a couple of days, which rapidly rearranges the air mass boundaries in different locations.

11.5. EASTERLY WAVES

An easterly wave is a migratory wave-like disturbance in the tropical easterlies. It moves from east to west, having a cycle of about 3 to 4 days and a wavelength of about 1200 to 1500 miles. They often bring rain showers and in some instances may develop into tropical cyclones. The southern countries of western Africa are a major source of these waves — about 60 waves are generated there every year. These move westward into the Atlantic and may affect weather systems even as far west as the eastern Pacific Ocean.

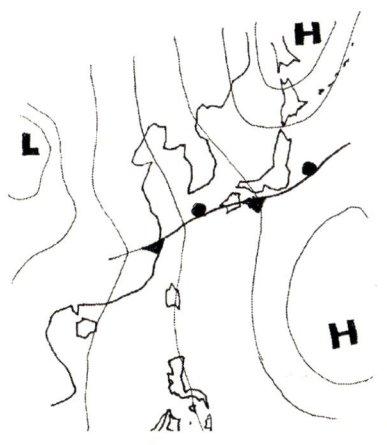

Figure 11-2. Typical appearance of Asia's subtropical stationary front in May. It occurs at the boundary of the building subtropical Pacific high and the receding Russian polar high, gradually moving north over the next few weeks. When visiting Japan in May, bring an umbrella!

11.5.1. Structure. The maximum intensity is usually in the lower and middle troposphere. Convective activity generally parallels the low level flow.

11.5.2. Cause. Intense heating in the Sahara desert produces a semipermanent warm-core barotropic low. This is reflected aloft as a high pressure area. In turn, an easterly upper jet sets up south of this low. Minor baroclinicity (horizontal thermal instability) within this jet then causes easterly waves to develop. This wave moves westward with the trade winds.

11.5.3. Types of easterly waves. Three different types of easterly waves have been identified, according to the slope of the trough with height.

* *Stable wave*. This is the most common type. The axis slopes to the east with height, and the wave moves slower than the prevailing flow, creating convergence to its east. Therefore most precipitation is found east of a stable wave. Strongest winds are usually near the surface.

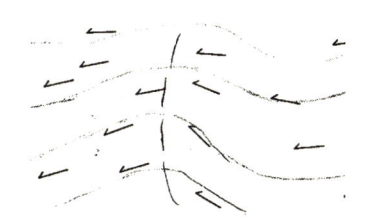

Figure 11-3. Typical appearance of easterly wave. This map uses wind plots and streamlines.

* *Neutral wave*. The axis is nearly vertical with height. The wave moves with the prevailing flow, and precipitation is more or less centered on it.

* *Unstable wave.* The axis slopes to the west with height. It moves faster than the prevailing flow, producing convergence along its leading edge. Therefore precipitation in an unstable wave is usually to its west. Winds are usually stronger aloft than at the surface.

11.6. MID-TROPOSPHERIC CYCLONES

These are cold-core barotropic lows which are caused either when a cutoff low becomes detached from the upper-level flow in middle latitudes or when an occluded low moves into tropical latitudes and becomes cut off from the mid-latitude westerlies. Most rain occurs on the eastern side of the cyclone, extending from 200 to 500 miles from the center, and the low may become stationary or drift erratically for days or even weeks. A good example of a mid-tropospheric cyclone is the "Kona" storm which is occasionally seen in the Hawaiian Islands in the winter.

11.7. TUTT

Occasionally the 300 mb or 200 mb analysis reveals a trough in the upper troposphere between subtropical ridges, particularly in the western parts of an ocean basin. The trough is often oriented SW-NE (Northern Hemisphere). These are referred to as TUTTs, or "tropical upper tropospheric troughs".

11.7.1. Impact. A TUTT usually contains little significant weather. It may enhance the formation of surface disturbances, and in the Pacific it is considered as a source of tropical cyclone development.

11.7.2. TUTT lows. A TUTT may occasionally "close off", producing a TUTT low. These may move erratically or remain stationary. As the TUTT low increases in size, it often moves west-southwestward at about 10 kts. During tropical cyclone season, TUTT lows are important because they represent a source of upper-level shear, which can disrupt the balance of circulation within tropical cyclones.

11.8. TROPICAL CYCLONES

A tropical cyclone is a warm-core barotropic low which has intensified due to widespread release of latent heat. It is a

---

Neutercane?
Back in the 1970's, the term neutercane was used to describe a small subtropical hurricane. The term was coined by the National Hurricane Center in 1971 as a result of studies based on satellite photographs. A neutercane typically develops between 25° and 35° along the Atlantic Coast, has a diameter of less than 100 miles, and has the option of developing into either a hurricane or an extratropical (frontal) cyclone. The term is no longer used.

Figure 11-4. Air Force WC-130 aircraft are routinely used to disperse dropsonde packages into the hurricane to measure pressure and other parameters. *(Courtesy NOAA)*

generic term for the regional designations "hurricane" and "typhoon".

11.8.1. Requirements. A tropical cyclone requires all of the following elements in order to develop and sustain itself. Since it represents a delicate balance of flow and thermodynamics, a single unfavorable ingredient can weaken the storm or even cause it to dissipate.

* *Heat source*. Warm ocean waters in excess of 80 deg F with about 200 ft or greater depth are required.

* *An unstable atmosphere*. Without this, no convection would develop, and no latent heat could be released to power the storm.

* *Coriolis force*. The tropical cyclone must be at least 300 nm from the equator. This ensures a balance of winds that prevents the low pressure area from being quickly filled by inflow.

* *Weak vertical wind shear*. If the shear from the surface to the upper troposphere is less than 20 kts (23 mph, 10 m s^{-1}), the storm can maintain a proper balance of winds.

* *A pre-existing low-level disturbance*. This provides a circulation for the storm to draw upon. Usually this disturbance is an easterly wave, but it can also be a complex of thunderstorms.

11.8.2. Locations. Here are some of the common locations for tropical storms.

* *Western North Pacific*. This is the most active location on the planet for tropical storms. Here they are called "typhoons" and affect the Philippines, China, Hong Kong, Vietnam, and Japan.

SAFFIR-SIMPSON HURRICANE SCALE
This scale is in widespread use in the United States. Listed are the maximum 1-minute sustained wind speeds, the minimum surface pressure, the storm surge, and the damage indications.

Category 1: MINIMAL. 74-95 mph (64-82 kts); greater than 980 mb; 3-5 ft. Damage primarily to shrubbery, trees, foliage, and unanchored homes. No real damage to other structures. Some damage to poorly-constructed signs.

Category 2: MODERATE. 96-110 mph (83-95 kts); 979-965 mb; 6-8 ft. Considerable damage to shrubbery and tree foliage; some trees blown down. Major damage to exposed mobile homes. Extensive damage to poorly-constructed signs. Some damage to roofing materials of low buildings; some window and door damage. No major damage to buildings.

Category 3: EXTENSIVE. 111-130 mph (96-113 kts); 964-945 mb; 9-12 ft. Foliage torn from trees; large trees blown down. Practically all poorly-constructed signs blown down. Some damage to roofing materials of buildings; some wind and door damage. Some structural damage to small buildings.

Category 4: EXTREME. 131-155 mph (114-135 kts); 944-920 mb; 13-18 ft. Shrubs and trees blown down; all signs down. Extensive damage to roofing materials, windows, and doors. Complete failure of roofs on many small residences. Complete destruction of mobile homes.

Category 5: CATASTROPHIC. 156+ mph (136+ kts); less than 920 mb; 19+ ft. Shrubs and trees blown down; all signs down. Considerable damage to roofs of buildings. Complete failure of roofs on many residences and industrial buildings. Extensive shattering of glass in windows and doors. Some complete building failures.

Flying in the eye
Here was one of Nature's most spectacular displays. The eye was a vast coliseum of clouds, 40 miles in diameter, whose walls rose like galleries in a great opera house to a height of approximately 35,000 feet where the upper rim of the clouds was smoothly rounded off against a background of deep blue sky. The sea surface was obscured by a stratocumulus undercast except for two circular openings on the east and west sides of the eye, respectively. Clouds in the undercast layer were grouped in bands which spiralled cyclonically about each of these openings or clear spots, both of which were approximately five miles wide. This horizontal alignment of clouds suggested the possibility that two separate small eddy circulations were present within the eye envelope. In the geometric center of the eye the stratocumulus undercast bulged upward in a domelike fashion to a height of 8,000 feet. Light turbulence in the tops of this dome was comparable to that in ordinary ocean cumulus. The walls on the eye to the west side were steep, either vertical or overhanging, and had a soft stratiform appearance. On the east side however clouds were more of a cumuliform type with a hard cauliflowery appearance.

R.H. SIMPSON, 1952

---

* *North Atlantic Ocean*. Disturbances move westward from Africa and may intensify, affecting the Caribbean islands, Cuba, Central America, and the United States. The source of these disturbances is usually easterly waves departing the west African coast.

* *Eastern North Pacific*. Late in the season it is quite common for tropical cyclones to develop west of Mexico, moving west or northwestward. Occasionally they make landfall in northwest Mexico, spreading significant moisture into the southern U.S. On rare occasion they travel as far west as Hawaii (e.g. Hurricane Iniki, 1992).

* *North Indian Ocean*. This area is west of Indonesia and affects India, Bangladesh, and Burma. Death tolls are extremely high due to the high storm surges over flat coastal regions, and are vastly underreported by the American media.

* *South Indian Ocean*. These storms develop southeast of Africa, where storms affect Mozambique and Madagascar.

* *Southeast Indian Ocean*. Tropical cyclones may develop off the northwest coast of Australia and affect northern and western parts of the country.

* *South Pacific Ocean*. Tropical cyclones are quite common northeast of Australia towards Fiji and locations eastward.

* *Arabian Sea*. On rare occasion a tropical cyclone may develop west of India in the Arabian Sea and affect adjoining countries.

* *South Atlantic Ocean*. This ocean is unique in that no significant tropical cyclones have been recorded.

11.8.3. Tropical cyclone type designations. Tropical cyclones have historically been classified according to their sustained wind speed. The sustained wind speed is a 10-minute average, however the U.S. differs in using 1-minute averaging.

* *Tropical disturbance*. In this stage, sustained winds are less than 34 kts (39 mph, 17 m s^{-1}), with an open circulation (no closed isobars).

* *Tropical depression*. Sustained winds are less than 34 kts (39 mph, 17 m s^{-1}), but there is a closed circulation (closed isobars). The storm is usually assigned a number.

* *Tropical storm*. Sustained winds are at least 34 kts (39 mph, 17 m s^{-1}) but less than 64 kts (74 mph, 33 m s^{-1}), with a definite closed circulation. The storm is usually assigned a name.

* *Hurricane/typhoon/cyclone/severe tropical cyclone.* Sustained winds are at least 64 kts (74 mph, 33 m s⁻¹). The name "hurricane" is given to storms in the Atlantic and East Pacific; "typhoon" in the west Pacific; "cyclone" in the western and northern Indian Ocean; and "severe tropical cyclone" in the south Pacific and southeast Indian Ocean.

* *Supertyphoon.* This term is used by the U.S. Joint Typhoon Warning Center in Guam to designate a typhoon with 1-minute sustained winds of 130 kts (150 mph, 65 m s⁻¹) or greater.

11.8.4. Tropical cyclone intensity designations. In the United States, tropical cyclones are rated by their Saffir-Simpson scale rating (see chart). All Category 3, 4, and 5 storms are referred to as "major" hurricanes, and are responsible for 80% of all tropical cyclone damage in the United States.

11.8.5. Structure. The following structures are often seen in tropical cyclones.

* *Outer rain bands.* These are lines of convection that form the outer spirals of the storm. The contain mostly heavy rain and thunderstorms.

* *Spiral bands.* These are lines of convection that seem to flow with the inflow into the center of the storm. They contain heavy rain, thunderstorms, and strong winds.

* *Eyewall.* This is the innermost area of intense convection, often forming a tight ring. It contains the highest sustained wind speeds within the storm and the heaviest rain.

* *Double eyewall.* Very strong tropical cyclones may show a double eyewall, which is a concentric ring of convection within a larger eyewall. This occurs when the spiral bands concentrate and form an outer eyewall. The divergence aloft produced by the outer eyewall produces convergence aloft over the inner eyewall, which eventually causes it to dissipate. The outer eyewall then replaces the inner one. The tropical cyclone often weakens for a short time as this happens.

* *Eye.* Strong tropical cyclones often develop an eye, containing only clear air and some low clouds. It is not known exactly what causes the eye. One theory suggests that divergence aloft over the eyewall convection meets over the exact center of the storm,

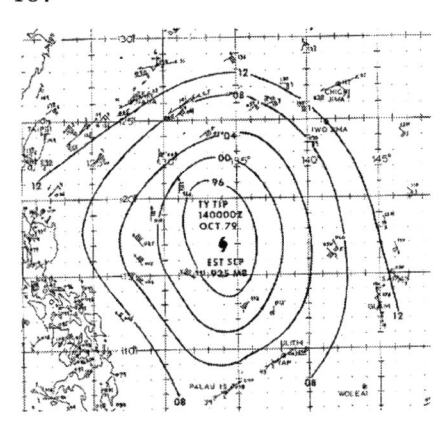

Figure 11-5. Supertyphoon Tip, the deepest typhoon in recorded history, showed a large pressure field that affected much of the western Pacific region. These isobars were drawn using ship reports only.

Typhoons get a facelift
Starting in 2000, English names for northwest Pacific typhoons such as Lois and Angela have been replaced by an internationally approved list of Asian names. Storms will now have names like Sarika and Megkhla, and many are named after flowers, animals, and even food. The Typhoon Committee of the U.N. Economic and Social Commission for Asia chose the 130 names from a list compiled by 14 member states, mostly in Asia. The United States contributed 10 names that are Palauan, Marshallese, and Chamorro in origin. The names will be operationally assigned by the Japanese Meteorological Agency in Tokyo, who has warning responsibility for the northwest Pacific region.

> **First hurricane encounter**
> The late 15th century brought the first hurricane encounter to European explorers. In 1494 Christopher Columbus made his second trip to the New World to establish a second Spanish settlement. It was named La Isabela and was founded on the northern coast of the Dominican Republic. Unfortunately within a couple of years two hurricanes hit and sank at least eight ships. In 1501 Peter Martyr wrote: "This strong south by southwest wind reached the city and the three ships that were alone at anchor. Without any perturbation of the water or surge of the sea, it broke their cables, gave them three or four twirls, and submerged them on the bottom." The losses coupled with diseases, parasites, and exposed anchorage led to the complete abandonment of La Isabela within four years.

> **Hurricane names**
> 2002: Arthur, Bertha, Cristobal, Dolly, Edouard, Fay, Gustav, Hanna, Isadore, Josephine, Kyle, Lili, Marco, Nana, Omar, Paloma, Rene, Sally, Teddy, Vicky, Wilfred
> 2003: Ana, Bill, Claudette, Danny, Erika, Fabian, Grace, Henri, Isabel, Juan, Kate, Larry, Mindy, Nicholas, Odette, Peter, Rose, Sam, Teresa, Victor, Wanda
> 2004: Alex, Bonnie, Charley, Danielle, Earl, Frances, Gaston, Hermine, Ivan, Jeanne, Karl, Lisa, Matthew, Nicole, Otto, Paula, Richard, Shary, Tomas, Virginie, Walter
> 2005: Arlene, Bret, Cindy, Dennis, Emily, Franklin, Gert, Harvey, Irene, Jose, Katrina, Lee, Maria, Nate, Ophelia, Philippe, Rita, Stan, Tammy, Vince, Wilma

creating convergence and producing downward motion. It is also thought that the storm's rotation centrifuges mass away from the center, causing low-level divergence which produces downward motion.

* *Outflow cirrus.* This is an area of high clouds, most often seen on infrared satellite imagery, that looks like a plume of cirrus expanding away from the storm. It often makes the storm appear larger than it really is. Outflow cirrus is produced by advection of moisture outward by upper-level divergence created within the storm as a whole.

11.8.6. Winds. The worst winds in a tropical cyclone are usually confined within the eyewall.

11.8.7. Precipitation. The tropical cyclone contains enormous amounts of precipitation, which may pose a serious flooding risk if the storm becomes stationary.

11.8.8. Hurricane-spawned tornadoes. Tornadoes occur in many hurricanes. Landfall is usually the catalyst for tornado production; as friction begins causing winds to diminish, these weaker winds coupled with the stronger winds a few thousand feet aloft creates strong low-level shear that favors tornado development. Since instability is fairly weak and low in altitude compared to typical tornadic storms, hurricane-spawned supercells usually are small, low-topped, and tend to occur during the daytime with the best insolation. This, coupled with the fact that they are rain-wrapped and move quickly makes detection and warning very difficult. A climatological study of hurricane-produced tornadoes found that they usually occurred on the outermost bands (50 to 200 miles from the center) ahead of and to the right of the storm's movement.

In September 1967 Hurricane Beulah produced 141 tornadoes in southeast Texas, most of them small and short-lived. Hurricane Gilbert in September 1988 brought several large tornadoes in south Texas which damaged parts of western San Antonio and were videotaped by storm chasers near Del Rio. In August 1992 Hurricane Andrew produced 62 tornadoes.

REVIEW QUESTIONS

1. Can tropical weather best be described as baroclinic or barotropic?

2. It is May and satellite imagery shows an east-west area of thunderstorms along the 10 deg N parallel, extending for over a thousand miles. What is this feature? What direction will it move over the weeks ahead?

3. Why are brisk trade winds often associated with fair weather?

4. What are the causes of easterly waves that later affect the Caribbean region?

5. Describe the structure and characteristics of a typical easterly wave.

6. You arrive in Hawaii for a vacation in February, and you find that heavy rains have been plaguing the islands for the past week. What tropical weather feature might be responsible?

7. Name all five critical ingredients required for tropical cyclone formation.

8. What is the difference between a hurricane and a typhoon?

9. Looking at satellite imagery, what feature gives the impression that a tropical cyclone is larger than it actually is?

10. What is the best rule of thumb for anticipating hurricane-spawned tornadoes?

HUMOR BREAK — The "other" Saffir Simpson ratings

Well, we might as well deface the Saffir-Simpson hurricane intensity scale by inventing categories 6 through 10. Here they are . . .
1 — MINIMAL (74-95 mph)
2 — MODERATE (96-110 mph)
3 — EXTENSIVE (111-130 mph)
4 — EXTREME (131-155 mph)
5 — CATASTROPHIC (156-218 mph)
6 — SERIOUSLY CATASTROPHIC (219-385 mph). Thankfully nothing this strong has hit the U.S yet. The government usually erases them off satellite pictures to avoid causing panic.
7 — NASTY (386-553 mph). Skyscrapers are torn and hurled.
8 — TOTALLY NASTY (554-673 mph). The Great Bahama hurricane of October 14, 554 B.C. (carbon-12 dating)
9 — HELLACIOUS (674-792 mph). Clouds turn red and storm goes around world once every 12 hours (e.g. Jupiter's Great Red Spot).
10 — VACUUMCANE (793+ mph). Hurricane's central pressure causes deaths due to hypoxia. Sonic booms from wind. Moon gets pulled pretty close to the Earth.

TIM VASQUEZ
"So, Ya Like Weather?," 1988

National Weather Service AWIPS workstations. *(Tim Vasquez)*

12 NUMERICAL GUIDANCE

Imagine a large hall like a theatre, except that the circles and galleries go right round the space usually occupied by the stage. The walls of this chamber are painted to form a map of the globe . . . A myriad of [human] computers are at work upon the weather of the part of the map where each sits, but each [human] computer attends to only one equation or part of an equation . . . Four senior clerks in the central pulpit are collecting the future weather as fast as it is being computed, and dispatching it by pneumatic carrier to a quiet room.

LEWIS F. RICHARDSON, envisioning a forecasting factory
"Weather Prediction by Numerical Process," 1922

Although the equations for motion were worked out in the early 20th century, forming the basis for mathematical prediction of the weather, computer power was simply not available at the time. Researchers envisioned a large forecast center filled with mathematicians who worked as a team, subdividing parts of the atmospheric puzzle, making calculations with slide rules and paper, and reassembling the results to make a forecast. Though this idea was never put into effect, the development of the first operational computers in the late 1940s and 1950s allowed the first models of atmospheric motion to be implemented. These numerical weather predictions were developed in the 1950s, showing a rapid increase in accuracy during the 1960s, and refined during the decades since then. Nowadays numerical guidance forms a valuable tool for helping the forecaster to envision how the atmosphere will change.

12.1. PREDICTION PROCESS

In the United States, numerical models are run at a computer complex at the National Center for Environmental Prediction (NCEP) in Maryland. The models are run about twice a day (depending on the model) and takes about an hour or more to produce the results. These are distributed through datastreams, radiofax, and the Internet to end users. Let's now take a look at how a model run comes together.

12.1.1. Analysis. In the first step of the analysis, the computer sorts through weather observations and maps the temperature, moisture, and pressure fields to a series of geographical grids. Some sophisticated methods have been developed for doing this. One of the most favored techniques is called "optimum interpolation", a three-dimensional method which considers the quality of the data and statistical relationships among the variables. Most of the models heavily use "first guess" fields to help fill in data-sparse areas with values from previous model runs. This significantly improves the accuracy of the run.

12.1.2. Initialization. The computer model thinks that the atmosphere is following equations. Therefore the initialization process is designed to adjust the analysis to be compatible with the prediction equations. Mathematical noise is removed from the fields, which fine-tunes the initial fields. Without this step, the prediction equations would produce unexpected errors, causing the model to "blow up".

Meteorological cancer
Forecasters are relinquishing their meteorological input into the operational product going to the user. Forecasters are operating more as communicators and less as meteorologists. Since this practice is increasing slowly with time, it can be called "meteorological cancer". By this is meant that today's forecaster can, if he chooses, and many do, accept numerical prognoses and guidance, put this into words, and go home. Not once does he have to use his meteorological knowledge and experience. This type of practice is taking place more and more across the United States, and it will be made easier to do with [computer technology].

LEONARD W. SNELLMAN
"Operational Forecasting Using Automated Guidance," 1977

12.1.3. Prediction. The primitive equations of motion and other equations derived from them, along with empirical relationships for friction, evaporation, precipitation, and so forth, are applied to the initial fields. This produces a prediction for a few minutes into the future. The computer repeatedly runs these equations, stepping forward a few forecast minutes at a time and producing new data fields. This is done until the run has gone as far into the future as requirements dictate. Statistical forecasts tend to generate the forecast directly without these intermediate steps.

12.1.4. Post-Process. Once the equations for a particular time are solved, the model output is interpolated from the model coordinates to the display coordinates. It may also be filtered somewhat for cosmetic purposes. The end results are what we see on the model output charts.

12.2. CLASSES

Forecast models may be of one of two basic classes, or a combination of the two.

12.2.1. Dynamical models. Most models we refer to (including in this book) are of this type. They use observations to simulate the behavior of the entire atmosphere, and from that they produce a forecast.

12.2.2. Statistical models. These are frequently used in forecasting tropical cyclones. Information about the current weather system is compared to historical data ("analogs"). This yields information about possible behavior of the current storm. Information about the atmospheric conditions are largely ignored, but this type of model is much simpler and easier to develop than dynamical models.

12.2.3. Combination models. Also used in tropical cyclone forecasting, combination models combine dynamical and statistical models to predict the behavior of a weather system. They are useful in data-sparse regions where it is difficult to accurately provide information to a dynamical model.

12.3. DOMAINS

A domain describes the geographic area represented by a model. This domain can either be limited-area or can be global.

On the use of models

I read and listen to forecast discussions now where the total focus is on the models. The debate is mostly aimed at the absurd exercise of trying to decide which is the "model of the day". The fact is that there is essentially nothing in the way of systematic procedures to decide the question reliably of which model to believe on any given day, in spite of most forecasters spending an inordinate amount of time trying to do so. Rather than using their time to diagnose atmospheric processes, forecasters often waste hours of their precious time in a futile effort to choose a model in which to believe. Models definitely have a role in science. That has never been a doubt, in my mind. What is troubling is the current overemphasis on modeling and the lack of interest evident in data and diagnosis of observations. The situation represents an imbalance among the triad of components in a healthy science: (1) theory, (2) observations, and (3) modeling. It has been discussed elsewhere that our science makes its most rapid advances when all these components of a successful science are in balance.

CHUCK DOSWELL
"On The Use of Models", 2000

NUMERICAL GUIDANCE • 163

12.3.1. Limited Domain Models. In the limited-domain model, the computer only considers a particular region of the Earth (such as within a box of North America). The problem with these domains is that after a certain amount of time, weather systems outside the domain of the model begin creeping in across the borders and affecting the weather inside the domain. Internal inaccuracies also develop. Reliable forecasts beyond 48 hours are often not possible with continental-scale limited domains.

12.3.2. Global Domain Models. Global domain models take into account weather across the entire Earth, so boundary errors cannot occur.

12.4. CONFIGURATION

Data within the model can be represented in one of two formats.

12.4.1. Gridpoint models. This is the simplest type of model. Data is mapped to an array of gridpoints, and model equations are applied to these gridpoints to produce a forecast map.

12.4.2. Spectral models. Data for the domain is transformed into a series of mathematical waves. Model equations are applied to these waves, and then this configuration is transformed back to geographic coordinates to produce a forecast map.

12.5. MODEL TYPES

There are various types of models that are in national or international use, which are described here. Many other types of models exist, but in general they are available only to a limited number of users and may never be encountered in day-to-day forecasting.

12.5.1. LFM Model. Also referred to as the Limited Area Fine Mesh, ERL (early look). (Finite differencing, limited domain, 160 km resolution, 7 layers). **This model has been discontinued by NCEP, but is included here because of its historical significance.** Large-scale hydrostatic meteorological equations are solved using finite differencing on a limited area covering most of North America and the adjacent oceans. The wind and mass variables are carried at all gridpoints, and fourth-order approximations are used for spatial derivatives. A modified form

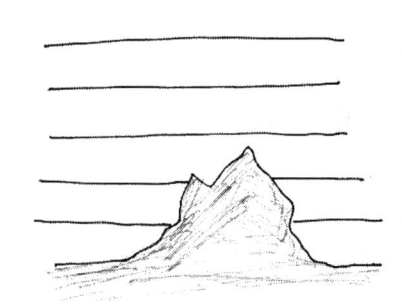

Figure 12-1a. Layers derived from pressure coordinates are fairly flat, following thermal patterns, and may intersect mountains and terrain. This can cause lots of problems in computer models.

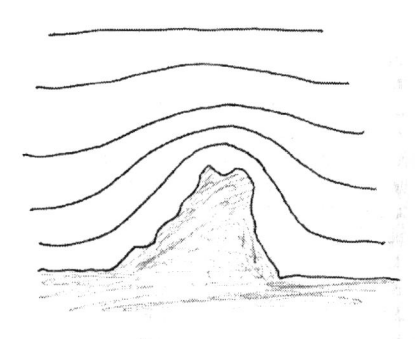

Figure 12-1b. Layers using eta and sigma coordinates bend with the terrain and never intersect it. The fundamental base of the sigma coordinate is at ground level, while the base of the eta coordinate is mean sea level. Most computer models use sigma coordinates, while the Eta model uses eta coordinates. The RUC uses a hybrid theta-sigma coordinate system, which is not described here.

Importance of map analysis
Last month I spent time evaluating the AWIPS 3-dimensional visualization tool. What I found to be most lacking was the absence of analysis of purely observational data, and I emphasized that over and over with the authors. If observational datasets can be introduced into the 3-D realm of visualization, I firmly believe that we will all be knocked on our heels with what we are seeing. I have some amazing images of the model's failures for May 3 [1999]! If we continue to rely on models and model initialization in place of "real" analysis, then I fear that Leonard Snellman's prediction of meteorological cancer will doom operational forecasting as we know it. Have we sold our science to the modelers and computer geeks?

Anonymous

---

of leapfrog, centered time integration is used. Lateral boundary conditions are obtained from the previous cycle forecast of the global model (AVN). Subgrid-scale physical processes (radiation, condensation, convection, and turbulent exchange) are parameterized in terms of grid-scale variables. The initial state is derived from an analysis of available observations on the model's grid. The analysis is performed using the successive correction method. Forecasts were made out to 48 hours twice a day.

12.5.2. Nested Grid Model (NGM), Regional Run (RAFS). (Finite differencing, 80 km grid meshed in a larger 160 km analysis grid, 16 sigma layers, optimum interpolation analysis). Large-scale hydrostatic equations are solved using finite differencing on a hemispheric domain. The wind and mass variables are staggered and a two-step forward time integration method is used. Lateral boundary conditions are based on an assumption of symmetry across the Equator. Initial conditions are produced by a Regional Data Assimilation System (RDAS) that incorporates all observations, including wind profilers and inflight ARINC Communication Addressing and Reporting System (ACARS) to update the NGM forecast at three-hour intervals over the 12 hours preceding the initial time of the current forecast. Subgrid-scale physical processes (radiation, condensation, convection, and turbulent mixing) are parameterized in terms of grid scale variables. Forecasts to 48 hours are made twice a day. The analysis creates 80 waves from the different fields before transposing them back to gridpoints. This gives a spherical-harmonic representation of the meteorological fields. The NGM is a vastly revised and improved version of the LFM. The NGM model contains a number of advanced physical parameterizations, such as vertical eddy transport, evaporation and heat transfer, longwave and shortwave radiation modelling, synoptic scale precipitation, and convective heating.

12.5.3. Eta Model (named after the Greek letter "eta", uses finite differencing, 30 km resolution, 30 eta layers). Large-scale hydrostatic meteorological equations are solved using the finite differencing method across North America. The time integration uses the splitting method in which gravity wave adjustment, advection, and physical processes are each solved with an appropriate time step. Lateral boundary conditions are required and ordinarily are obtained from the previous cycle of the AVN global model forecast. Small-scale physical processes not resolved by the model (radiation, condensation, convection, and turbulent exchange) are parameterized in terms of the resolved scale variables. The model is unique in its treatment of orography as a set of steps, each of which corresponds to a level of the model's vertical coordinate, which is referred to as eta. The initial state is derived in one of two ways: from an analysis in the model's vertical coordinate of all available observations using a first guess provided by the GDAS, or by interpolation of the regional analysis (RAFS/NGM) to the Eta model's grid. The model is usually run out to 48 hours twice daily.

12.5.4. Spectral (AVN), Medium Range Forecast (MRF), and Global Spectral Model (GSM)

(Spectral, global domain, 80 waves, 12 sigma layers). Large-scale hydrostatic equations are solved using the spectral method on the entire globe. The time integration uses the semi-implicit method; additionally, the semi-lagrangian approximation is used for horizontal advection. No lateral boundary conditions are required. Small-scale physical processes not resolved by the model (radiation, condensation, convection, and turbulent exchange) are parameterized in terms of the resolved scale variables on the gaussian grid of the model. The adjustment of model variables in response to the small scale physical forcing is projected spectrally onto the resolved spectrum of the model's dependent variables. The initial state is derived from an analysis of all available observations covering the globe. The analysis is performed using a spectral statistical interpolation method to update a first guess produced by the Global Data Assimilation System (GDAS). Forecasts producing global weather guidance to aviation and the general public are made twice each day. These forecasts cover 120 hours. Once each day, a medium range forecast (MRF) run is made out to 240 hours; it uses more observational data than the AVN/GSM runs owing to a much later data cutoff time. The spectral model was first introduced into NMC's operations in August 1980 in a 12-layer 30-wave rhomboidal configuration, but since then it has improved to 80 waves.

12.5.5. Quasi-Lagrangian Model (QLM).

Also known as the Hurricane Run (HCN) (gridpoint, 40 km resolution, 16 sigma layers, 4400 x 4400 km domain). This was implemented in 1988 to replace the MFM (movable fine-mesh model) that was used operationally by the National Hurricane Center. Large-scale hydrostatic equations are solved using finite differencing on a limited area domain. Lateral boundary conditions are obtained from the AVN global model. The model's name comes from its use of a lagrangian approximation for the time derivative following the motion on the air parcels. Subgrid-scale physical processes (condensation, convection, and turbulent transfer) are parameterized in terms of the grid scale variables. The hurricane is parameterized at the initial time using an idealized vortex and

---

In spite of progress made in the development of quantitative [computer] forecast techniques, the conventional forecaster will have an important part to play. His wide experience of local and regional conditions, orographic and topographic influences, moisture and pollution sources, etc, will be invaluable in supplementing the machine-made forecasts. While the machines provide the answers that can be computed routinely, the forecaster will have the opportunity to concentrate on the problems which can be solved only by resort to scientific insight and experience. Furthermore, since the machine-made forecasts are derived, at least in part, from idealized models, there will always be an unexplained residual which invites study. It is important, therefore, that the forecaster be conversant with the underlying theories, assumptions, and models. In particular, it is important that he be able to identify the "abnormal situation" when the idealized models (be they dynamical or statistical) are likely to be inadequate.

SVERRE PETTERSSEN,
"Weather Analysis and Forecasting," 1956

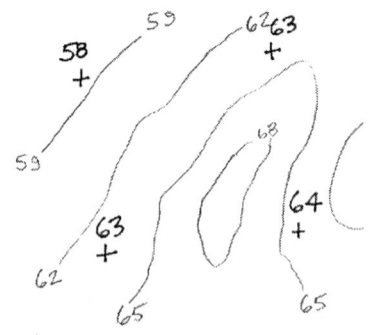

Figure 12-2. Mesoscale features can easily be lost or smoothed over in forecast models. In the example above, the isotherms assume we "see all, know all" in the atmosphere and a tongue of high dewpoints (near 70 degrees) exists. Unfortunately, it might not be picked up by the observation network or might be smoothed during initialization, and the equations would only see the accurate but coarsely-sampled values shown at the gridpoints. The forecast output could end up being significantly erroneous.

information is provided by the National Hurricane Center (NHC). The model is usually run during the hurricane season when requested by NHC. Forecasts of storm tracks are made to 72 hours and transmitted to NHC as guidance for their official forecasts.

12.5.6. ECMWF Model. Referred to as the ECMWF Model, European Center for Medium-Range Weather Forecasts Model, European model (spectral, global domain, 243-waves (55 km), and 31 sigma layers). It is run once a day on 1200 UTC data. The ECMWF model has a high-quality package of physical parameterizations that handle convective transport, condensation and evaporation, radiation, and other processes. Its forecasts are often very accurate for North America. The ECMWF model is similar to the AVN/GSM model in that it solves the same fundamental set of hydrostatic equations, however it differs significantly in its parameterizations of subgrid-scale physical processes (radiation, convection, and turbulent transfer). Furthermore the databases used in establishing the initial conditions for the numerical integration differ due to differences in methods of data assimilation and the cut-off time for initiating the forecasts. When comparing American models to the ECMWF most forecasters don't consider the ECMWF's differences in physical parameterizations into account but instead consider the recent skill of each model in handling the weather situation, and to a lesser extent the differences in the definition of the initial conditions for the forecasts.

12.5.7. UKMET Model. The UKMET (United Kingdom Meteorological) model is a finite-differencing model with a resolution of 70 km (1.25 deg longitude by 0.83 deg latitude spacing) and 20 layers. It is run at the United Kingdom Meteorological Office in Bracknell, England. It is used mostly for 3 to 5 day forecasting guidance.

12.6. LIMITATIONS OF MODELS

It is extremely important to remember that models are only simulations of the atmosphere using current data. They are never 100% correct, and in unusual situations (e.g. major storms) significant errors may be seen. Here are some factors that contribute to the inaccuracy of numerical models.

12.6.1. Small amount of observational data. A model might have great resolution — 30 or 40 miles, but useful observations may be 200 to 300 miles apart. Assuming the observations are accurate, we can say that the more inaccurate the initial analysis is, the greater the likelihood that forecast charts will contain inaccuracies, especially as the computer steps forward in time.

12.6.2. Boundary errors. A global-domain model such as the GSM (AVN/MRF) is immune to boundary errors since the model encompasses all points of the globe. However, limited-domain models such as the NGM and ETA only cover a limited area. Therefore the edges of the domain gradually become unrepresentative with time (the model does not account for real-world systems outside the boundary of the model). On the ETA run, for example, significant errors are seen in the central Pacific at the 48 hour point due to boundary errors.

12.6.3. Loss of mesoscale features. Grid spacing often precludes detection and prediction of some subsynoptic features. There are many short waves in the lower and middle troposphere that are only a few degrees or less wide in longitude. Since each one is baroclinic, each short wave has its own area of upward and downward vertical motion. The model may be unable to forecast it and the weather it creates.

12.6.4. Inadequate parameterization of physical processes. Processes such as solar radiation, terrestrial radiation (including radiation reflected and re-reflected by cloud tops, bases, etc), evaporation from below clouds, precipitation, and other processes must be explicity or implicitly accounted for in order to have the model produce a realistic simulation of the atmosphere. Clouds and precipitation are some of the most important physical processes a model must predict. Unfortunately these processes may not be adequately accounted for in the model, or in simple models, some processes may be ignored.

12.6.5. Parameterization of convection. Most operational forecast models handle convection unrealistically. There is inadequate computer power to resolve these processes, many of which are not fully understood. Furthermore, if a major thunderstorm is occurring at the time of a sounding and this sounding is inadvertantly entered into the analysis of a model, the model could blow the situation up into a hurricane-like storm (this is called "convective feedback"). The forecast models of the Air Force and Navy are aimed more at cloud prediction because of obvious aviation concerns, therefore they include stronger attempts at parameterizing for convection.

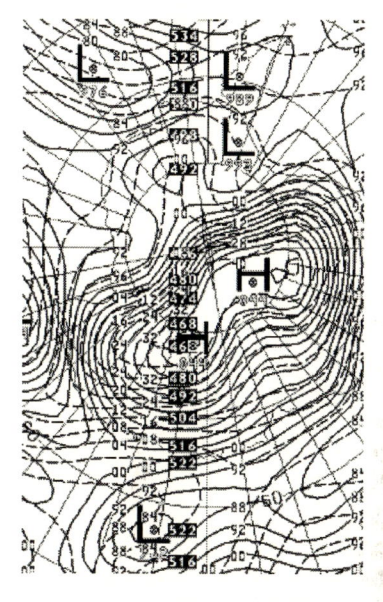

Figure 12-3. Yes, thanks to the AVN/MRF model we do have forecast coverage of the world, even of Antarctica and surrounding waters (seen above for a day in April 2000). However the southern hemisphere has a substantial lack of weather data, and it is frequently not even known whether the model initializations are representative. It is a strong example of when to be wary of using computer guidance.

REVIEW QUESTIONS

1. Why is numerical modelling such a new science?

2. Describe what would likely happen if a numerical model run is not properly initialized.

3. There is a massive telecommunications outage and weather observations for Canada have been unavailable for the past day. However you take a look at the 00-hour ETA panels and determine that it accurately depicts conditions in Canada. How did the model know what the conditions were in Canada?

4. The NGM model is a limited domain model. Why does this make it useless for forecasting weather 3 days in advance?

5. Name one numerical model used in the United States that incorporates a global domain.

6. A slow-moving cold front moves into Texas from Mexico, and was completely unforecast by the computer models. What is the most likely culprit?

7. A large snowstorm moves through the northern United States, and for the next ten days forecasters in Illinois, Iowa, and Wisconsin notice degraded numerical model performance. What is the most likely culprit.

8. Name a reason that a typical numerical model is unable to accurately forecast the occurrence of an individual thunderstorm.

9. The National Center for Environmental Prediction (NCEP) develops a new model that makes forecasts solely based on comparisions with historical data. What type of model is this?

10. What are the characteristics of a spectral model?

APPENDIX

CODES QUICK-REFERENCE

This is a quick reference guide to some of the most commonly-used weather reporting formats. For other formats, download our codes section from our web site (see the appendix).

Surface METAR observation format

METAR is the most commonly-encountered format used for surface observations and is used heavily within the United States.

```
KICT 161553Z AUTO 35008G17KT 1 1/2SM -RA BR BKN011 OVC016 21/20
A3005
     RMK AO2 RAB02 SLP169=
```

KICT — Station identifier for Wichita Kansas. To search or get a list of identifiers, visit the NWS's station search page on the Internet at:
http://www.nws.noaa.gov/oso/siteloc.shtml

161553Z — The 16th day of the month at 1553 GMT (GMT is also known as "UTC", "Z", or "zulu" time). Eastern Time is 5 hours behind GMT time (4 hours when on daylight saving time).

AUTO — An automated station prepared the report. Some automated stations are augmented by a human.

35008G17KT — Winds are blowing from 350 deg (north) at 8 knots gusting to 17 knots (KT). Winds are always referenced to true north, not magnetic north.

1 1/2SM — Visibility is 1.5 statute miles. European stations typically report in meters or kilometers.

-RA BR — Light rain (-RA) and mist (BR) is occurring. Other symbols that may be seen are TS (thunderstorm), FG (fog), SN (snow), SH (showers), GR (hail), FZ (freezing precipitation), DZ (drizzle), and others. A minus prefix means light; a plus prefix means heavy.

BKN011 OVC016 — A broken layer (60 to 90% cover) of clouds exists at 1100 ft above ground level. A second layer of clouds, overcast (100% cover) exists at 1600 ft above ground level. A layer can be CLR or SKC (clear, 0%), FEW (less than 10%), SCT (scattered, 10-50%), BKN (broken, 60-90%), or OVC (overcast, 100%).

21/20 — Temperature and dewpoint is 21 and 20 deg C, respectively. A prefix of M means "minus".

A3005 — Altimeter setting (pressure) expressed in hundreds of inches. In this example the value is 30.05 inches of mercury.

RMK — Indicates remarks follow.

AO2 — The type of automated station is AO2 (ASOS).

RAB02 — Rain began at 2 minutes past the hour.

SLP169 — The sea-level pressure is 1016.9 millibars. If SLP is above 500 to 600, a 9 rather than 10 is prefixed, which indicates a value in the 900s range.

Surface SYNOP observation format

```
72365 11966 82504 10074 21001 39875 40157 52008 69901 70206 8807/
```

72365 — WMO Index number for Albuquerque, NM.
11966 — Coded visibility and cloud height values (will not be decoded here).
82504 — Total cloud cover in eighths is "8", winds are from 250 ("25") at 4 ("04") knots.
10074 — "1" is marker for temperature, "0" is sign (positive), and temperature is "074" (7.4 deg C).
21001 — "2" is marker for dewpoint, "1" is sign (negative), and dewpoint is "001" (-0.1 deg C).
39875 — "3" is marker for station pressure data (will not be decoded here).
40157 — "4" is marker for sea level pressure data, "0157" is pressure in mb in tens (if under 5000 add 1000 mb); in this case the pressure is 1015.7 mb.
52008 — "5" is marker for pressure change data. "2" is change type; "008" is change in hundreds of mb.
69901 — "6" is marker for rainfall data (will not be decoded here).
70206 — "7" is marker for weather occurrences; "02" (no significant weather) was occurring now; types "0" (none) and "6" (rain) had occurred during the past observation period.
8807/ — "8" is marker for cloud data. Height of lowest cloud was "8" (will not be decoded here), lowest cloud was "8" (stratocumulus), middle cloud was "7" (altocumulus), and high cloud was "/" (not observable).

Upper air (35-X TEMP) RAOB format

The 35-X TEMP format is the most commonly-encountered format used for upper air observations throughout the world. For detailed information, please see the references.

All reports start with the following information:
70026 — WMO index number: Barrow, Alaska.
TTAA/TTBB/PPBB — Indicates that mandatory level temperatures (TTAA), significant level temperatures (TTBB), or significant level winds (PPBB) follow.
66121 — The "66" is the date of the month; 50 is added when the wind speeds are in knots, otherwise winds are in meters per second. In this case, it is the 16th and winds are in knots. "12" is the GMT hour, in this case, 1200 GMT. The "1" is a code that indicates what kind of data is used in the report (not described here).
70026 — The WMO index number is repeated.

```
70026 TTAA  66121 70026 99010 05006 10006 00090 09606 10506
92745 12660 19005 85447 06636 20008 70012 02356 23011 50561
18564 20021 40723 29164 19529 30922 44959 21025 25041 55158
22524 20184 49159 21024 15375 46564 19015 10644 46165 16509
88231 57958 21031 77999 51515 10164 00005 10194 17506 22009=
```

70026 TTAA 66121 70026 — See notes above
99010 — The surface ("99") level has a pressure figure of "010" (decoding will not be provided here).

05006 — The temperature is 05.0 deg C (if the tenths digit is odd, it signifies a negative temperature). The dewpoint depression "06" indicates the dewpoint is 0.6 C degrees lower than the temperature. If the value exceeds "50" then subtract 50 to get the value in whole degrees (e.g. "58" means 8 C deg).
10006 — Winds at this level are 100 true at 06 knots.
The above 3 groups are repeated over and over.

```
70026 TTBB  66120 70026 00010 05006 11005 05807 22000 09606
33993 12001 44963 13219 55954 13256 66925 12660 77841 05823
88788 02637 99748 01108 11719 00957 22610 09538 33586 10750
44579 10960 55572 11357 66555 12760 77543 13568 88531 14960
99499 18764 11493 19556 22484 20927 33468 22924 44462 22962
55453 22567 66346 37358 77231 57958 88216 55758 99202 48959
11168 45564 22100 46165 31313 01102 81102=
```

70026 TTBB 66120 70026 — See notes above
00010 — "00" is a sequence number, which increases to 11, 22, 33, 44, etc. It's there merely to help quickly identify each group visually. The "010" is the level in millibars for the following data (if it's less than 100, add it to 1000; in this case we have 1010 mb).
05006 — The temperature at this level is 05.0 deg C and dewpoint depression is 0.6 deg C, yielding a dewpoint of 4.4 deg C. See rules above under TTAA.
The above 2 groups are repeated over and over.

```
PPBB  66120 70026 90012 10006 13504 18505 90346 19505 20007
20509 90789 22009 23008 23008 91024 23011 22016 20022 9156/
19525 20025 92035 20021 19530 20028 93035 21026 24024 21025
9447/ 19014 20012 9503/ 18009 17010=
```

PPBB 66120 70026 — See notes above.
90012 — This group indicates the height of the following data. "9" is a placeholder. "0" is the height in ten-thousands of feet. "0", "1", and "2" indicates the height in thousands of feet of the first, second, and third group respectively. So the groups to follow are for 0, 1000, and 2000 ft MSL.
10006 — Winds at 0 ft are 100 at 6 kts.
13504 — Winds at 1000 ft are 135 at 4 kts.
18505 — Winds at 2000 ft are 185 at 5 kts.
These groups repeat as necessary.

READING

Barry, R. G. and Chorley, R. J., 1998: <u>Atmosphere, Weather & Climate</u>, Methuen & Co, Ltd, London. ISBN 0-415-16020-0. In print.

Djuric, D., 1994: <u>Weather Analysis</u>, Prentice Hall, Englewood Cliffs, NJ. ISBN 0-135-01149-3. An excellent and highly detailed primer for all enthusiasts and meteorologists. Facsimile version currently in print.

Doswell, C. A., 1982: <u>The Operational Meteorology of Convective Weather</u>: Volume 1, Operational Mesoanalysis, National Weather Service, Washington, D.C. Out of print but available through National Technical Information Service (www.ntis.gov) and for sale in PDF format at weathergraphics.com.

Gedzelman, Stanley David, 1980: <u>The Science and Wonders of the Atmosphere</u>, John Wiley & Sons, New York. ISBN 0-471-02972-6. Out of print but excellent.

Kurz, Manfred, 1990: <u>Synoptic Meteorology</u>, Deutscher Wetterdienst, Offenbach, Germany. ISBN 3-88148-338-1. Availability unknown.

Moore, J. T., 1992: <u>Isentropic Analysis and Interpretation</u>, National Weather Service Training Center, Kansas City, MO. Internal forecaster handout, not generally available except perhaps in local National Weather Service office libraries.

National Weather Service, 1993: <u>Forecaster's Handbook #1</u>, Washington, DC. This is mainly an overview of NWS products, not a forecasting how-to. Source of this document is the National Weather Service.

Nielsen-Gammon, J. W., 1995: <u>Introduction to Isentropic Analysis</u>, Texas A & M University, College Station, TX. Probably available through Texas A&M meteorology department.

Office for the Federal Coordinator of Meteorology, 2000: <u>Federal Meteorological Handbook #1</u>: Surface Weather Observations and Reports. Available in its entirety online at http://www.ofcm.gov/fmh-1/cover.htm . This describes the complete set of practices for taking and coding observations in the U.S.

Saucier, W. J., 1983: <u>Principles of Meteorological Analysis</u>, Dover Publications, New York. ISBN 0-486-65979-8. This is a reprint of an older (circa 1960s) title. Provides a complex theoretical framework for meteorological analysis. Limited quantities were still available through Amazon.com in November 2000.

World Meteorological Organization, 1995: <u>Publication 306: Manual on Codes, Volume 1, Part A: Alphanumeric Codes</u>. WMO, Geneva. ISBN 92-63-15306-X. A complete guide to the international coding formats for METAR, upper air, and dozens of other obscure types. Unfortunately it comes with a high price tag; be prepared to pay well over $100.

SOFTWARE

Although the Internet itself provides a valuable source of data, most experienced forecasters realize that the Web does not provide rapid methods for analyzing, plotting, and dissecting data. Fortunately there are several powerful programs that accomplish this.

DIGITAL ATMOSPHERE. This unique software program, designed for Windows 95/98/2000 and NT, allows flexible plotting of surface and upper charts, soundings, and climatology. It even performs 3-D roaming soundings and cross sections. Coverage is worldwide. Automatically retrieves free weather data on its own via the Internet. Digital Atmosphere can be downloaded from:

```
http://www.weathergraphics.com/da/
```

RAOB. Although it is still a MS-DOS program as of November, 2000 it is still the most extensive and detailed software available for sounding and cross section analysis. A demo of RAOB can be downloaded from:

```
http://www.weathergraphics.com/raob/
```

EDUCATIONAL WEB SITES

METEOROLOGY GUIDE. This is one of the best online weather education sites. It leans toward the basics, but there is some interesting information on weather analysis.
http://ww2010.atmos.uiuc.edu/(Gh)/guides/mtr/home.rxml

JOSEPH BARTLO'S ARTICLES. It's surprisingly tough to find good articles about weather analysis and forecasting on the Internet. But Joseph Bartlo has a gold mine of them. Check them out.
http://www.enter.net/~jbartlo/assortment.htm

NWS METAR PAGE. Great information on the METAR and TAF formats, as well as a station ID guide.
http://www.nws.noaa.gov/oso/oso1/oso12/metar.htm

CHUCK DOSWELL'S ARTICLES. Some great articles from a leading retired National Oceanic and Atmospheric Administration researcher and an expert weather analyst:
http://www.nssl.noaa.gov/~doswell/Essays_index.html
http://www.nssl.noaa.gov/~doswell/Web_Formalpubs.html
http://www.nssl.noaa.gov/~doswell/inforpub.html
http://webserv.chatsystems.com/~doswell/opfun.html

ROGER EDWARDS' DATA LINKS. A great place to start to see Web-based graphics useful for forecasting purposes.
http://www.stormeyes.org/tornado/rogersif.htm

IWIN. All the National Weather Service bulletins you could ever want. Great for monitoring weather events.
http://iwin.nws.noaa.gov/

FORECAST HANDBOOK. This is the official web site for the Forecast Handbook. You can download a free section on meteorological codes. You can also download lots of weather software and look through information and products relating to weather forecasting and analysis.
http://www.weathergraphics.com/

THE FAQ FOR HURRICANES, TYPHOONS, AND TROPICAL CYCLONES. This resource by Christopher W. Landsea is the definitive starting point for finding out more about tropical storms.
http://www.aoml.noaa.gov/hrd/tcfaq/tcfaqHED.html

THERMODYNAMIC DIAGRAMS. An excellent historical source for learning about thermodynamic diagrams.
http://www.booty.demon.co.uk/metinfo/thdyndia.htm

GOVERNMENT WEATHER AGENCY WEB SITES

Following are web site addresses of official government weather agencies. If you get a "404 Error" or the link no longer works, you can often try shortening the URL by removing data that follows the "/" characters. For translation to English from any of several major languages, try the Babelfish tool at http://babelfish.altavista.com/ . For other sites visit the WMO link below.

Argentina	`http://www.meteofa.mil.ar/`
Australia	`http://www.bom.gov.au/`
Brazil	`http://www.inmet.gov.br/`
Canada	`http://www.msc.ec.gc.ca/`
China	`http://www.cma.gov.cn/`
Cuba	`http://www.met.inf.cu/`
France	`http://www.meteo.fr/`
Germany	`http://www.dwd.de/`
Italy	`http://www.meteo.difesa.it/`
Japan	`http://www.kishou.go.jp`
Korea	`http://www.kma.go.kr/`
Kenya	`http://www.meteo.go.ke/`
Mexico	`http://smn.cna.gob.mx/`
Netherlands	`http://www.knmi.nl/`
NZ	`http://www.met.co.nz/`
Norway	`http://www.dnmi.no/`
Pakistan	`http://met.gov.pk/`
Philippines	`http://www.pagasa.dost.gov.ph/`
Poland	`http://sunsite.icm.edu.pl/meteo`
Russia	`http://www.mecom.ru/`
Singapore	`http://www.gov.sg/metsin/`
Saudi Arabia	`http://www.mepa.org.sa/`
South Africa	`http://cirrus.sawb.gov.za/`
Spain	`http://www.inm.es/`
Sweden	`http://www.smhi.se/`
Switzerland	`http://www.meteoschweiz.ch/`
Taiwan	`http://www.cwb.gov.tw/`
Thailand	`http://www.thaimet.tmd.go.th/`
Turkey	`http://www.meteor.gov.tr/`
UK	`http://www.meto.govt.uk/`
USA	`http://www.nws.noaa.gov/`
WMO	`http://www.wmo.ch/`

Top ten weather forecasting myths

10. "MAMMATUS INDICATES TORNADOES." It's often seen on the weakest of storms, such as those in the tropics, and even on the underside of altostratus layers.

9. "UPPER COLD ADVECTION RESULTS IN DESTABILIZATION." Air follows isentropic surfaces, not constant pressure surfaces. Therefore the appearance of cold advection on a constant-pressure chart (such as 500 mb) does not necessarily mean cooling will occur downstream.

8. "TORNADO INTENSITY IS MEASURED WITH THE FUJITA SCALE." The Fujita scale is a *damage* scale, not an intensity scale.

7. "UPPER DIFLUENCE MEANS UPWARD MOTION." Difluence is frequently cancelled out by speed convergence.

6. "FALLING SNOW WILL MELT WHEN IT'S OVER 32°F." If the air is dry, any melting actually causes heat to be removed from the air, resulting in cooling.

5. "POSITIVE VORTICITY ADVECTION MEANS UPWARD MOTION." Seeing PVA at 500 mb only provides an *inference* of *one* term of the omega equation for vertical motion. It is frequently cancelled out by thermal advection.

4. "TROPICAL CYCLONES DIE AFTER MAKING LANDFALL BECAUSE OF FRICTION." Research has shown that factors such as the lack of moisture and heat sources are the real catalyst for dissipation. Ironically, the friction just after landfall actually increases turbulent flow and brings stronger winds down to the surface.

3. "HIGHER RESOLUTION MODELS MEAN BETTER FORECASTS." Operational models as fine as 30 km are now in use, and in recent years many weather agencies have been increasingly blinded by the emphasis on using pushbutton technology to perform forecasting. Unfortunately the vast increases in computing power and model resolution have brought diminishing returns. There has been little change since 1960: we are still sampling the atmosphere *using an average upper-air spacing of at least 600 km and a surface spacing of 100 km*, and calls in scientific journals to improve the radiosonde data network have largely fallen on deaf ears.

2. "THAT TORNADO OCCURRED BECAUSE OF EL NINO / LA NINA / GLOBAL WARMING, ETC." This is a fallacy that largely is attributed to uninformed journalists. A close analogy to this fallacy is, "My car got rear-ended because of the heavy traffic we had after the ball game." A climatological phenomena with a cycle of six months or more cannot possibly cause a ten-minute weather occurrence. There are indications that variations in the large-scale circulation may have some influence on the environment to favor severe thunderstorms in one area of the country over a long-term period, but in the end tornadoes are still caused by small-scale processes in the thunderstorm.

1. "TOILETS SPIN THE OTHER WAY IN THE SOUTHERN HEMISPHERE." OK, well maybe this isn't a *forecasting* myth, but it is definitely one of the most prominent myths regarding weather. It's no doubt that tourist traps on the Equator combined with publicity from poorly-researched travel television shows have done quite a bit to perpetuate this myth. Simple equations show that the contribution of the Coriolis force to such a small-scale process is infinitesimally small, and is vastly overpowered by the trajectory of water entering the bowl as conservation of angular momentum imparts spin to the fluid. In other words, you can pour a bucket of water into the toilet and control the spin in either direction. Therefore spin is a function of exactly how water enters the bowl according to the toilet's design. See:
    http://www.ems.psu.edu/~fraser/Bad/BadFAQ/index.html
for lots of great information on this topic.

Analysis Exercises

Provided in this section are a series of real-world charts for you to analyze. These aren't simply exercises for big "superstorm" days where the fronts and patterns jump off the page. These are tougher and more instructive: everyday weather systems with a variety of subtleties and strengths.

Your objective is to analyze for the following:

SURFACE CHARTS
— Isobars (every 2 mb)
— High and low pressure areas
— Fronts, if any
— Troughs and other boundaries

UPPER AIR CHARTS
— Contour interval as specified in "Tools" chapter
— Highs and lows
— Jet axes (500 mb or above)
— Short wave trough and ridge axes (500 mb)

There are no restrictions on photocopying the attached maps.

Suggested solutions are provided on the official Weather Forecasting Handbook site at:
    http://www.weathergraphics.com/fcstbook/

You can also download the original maps off the site if the maps in this book are illegible.

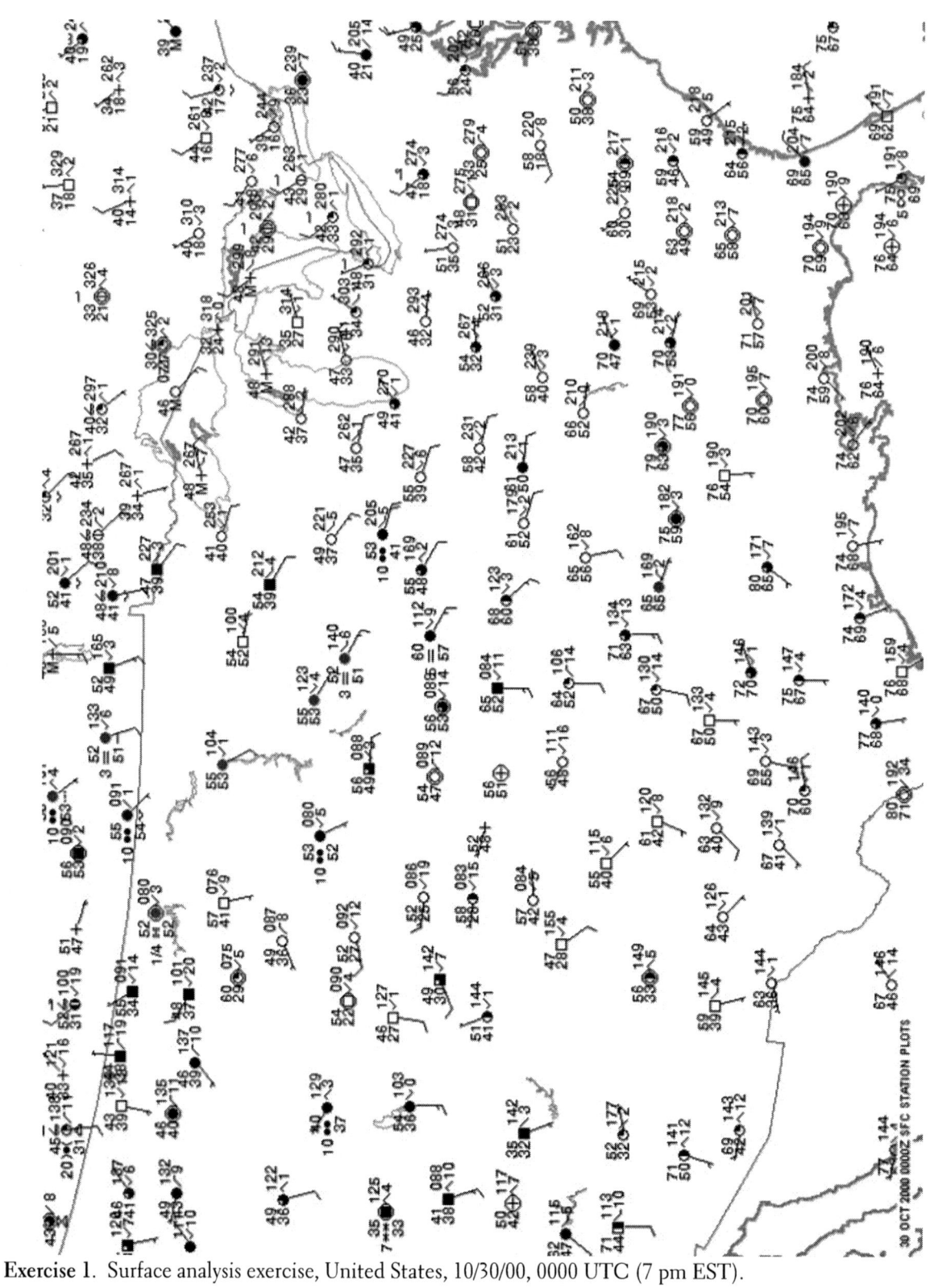

Exercise 1. Surface analysis exercise, United States, 10/30/00, 0000 UTC (7 pm EST).

ANALYSIS EXERCISES • 183

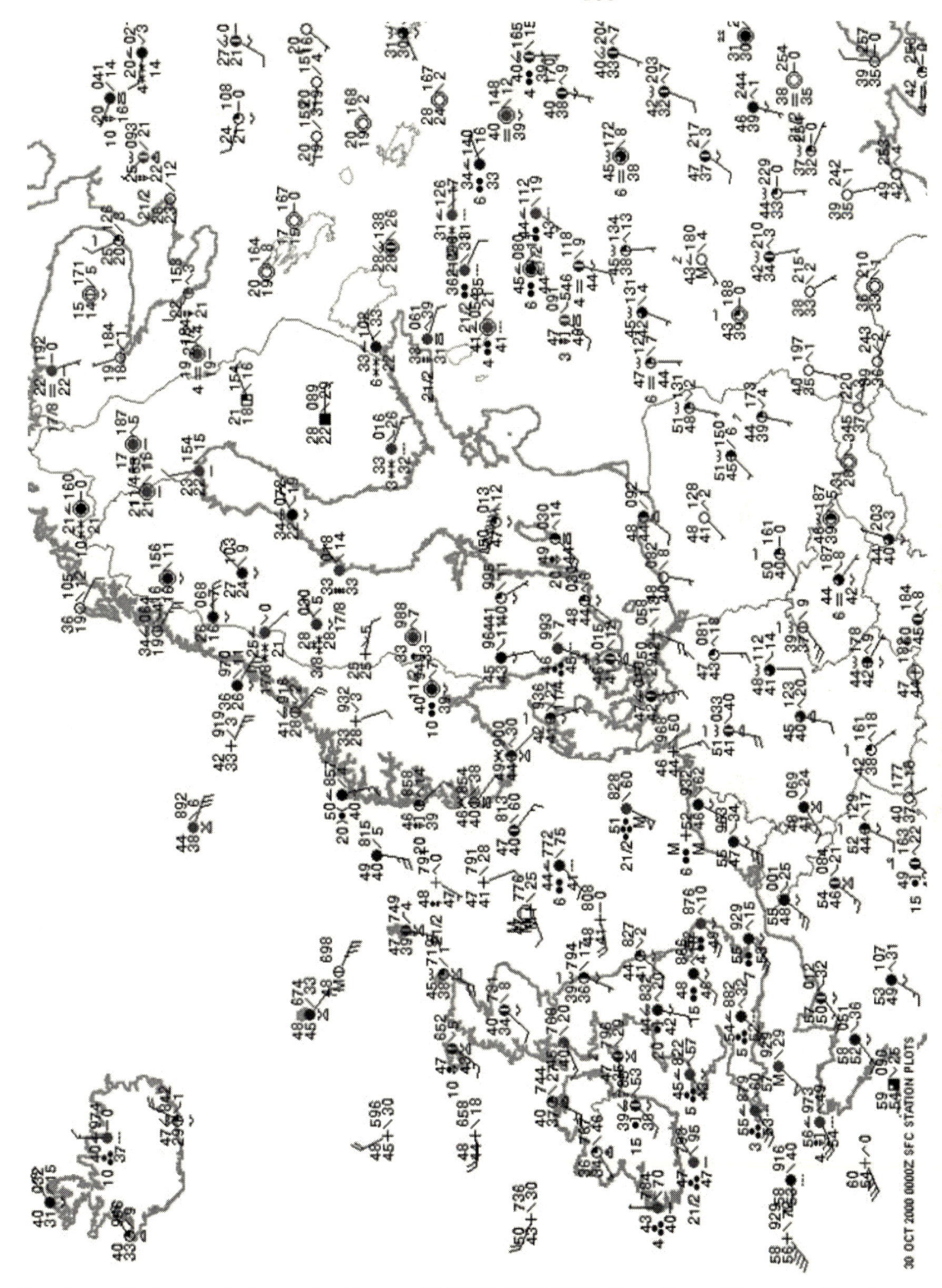

Exercise 2. Surface analysis exercise, Europe, 10/29/00, 0000 UTC (1 am Central European Time).

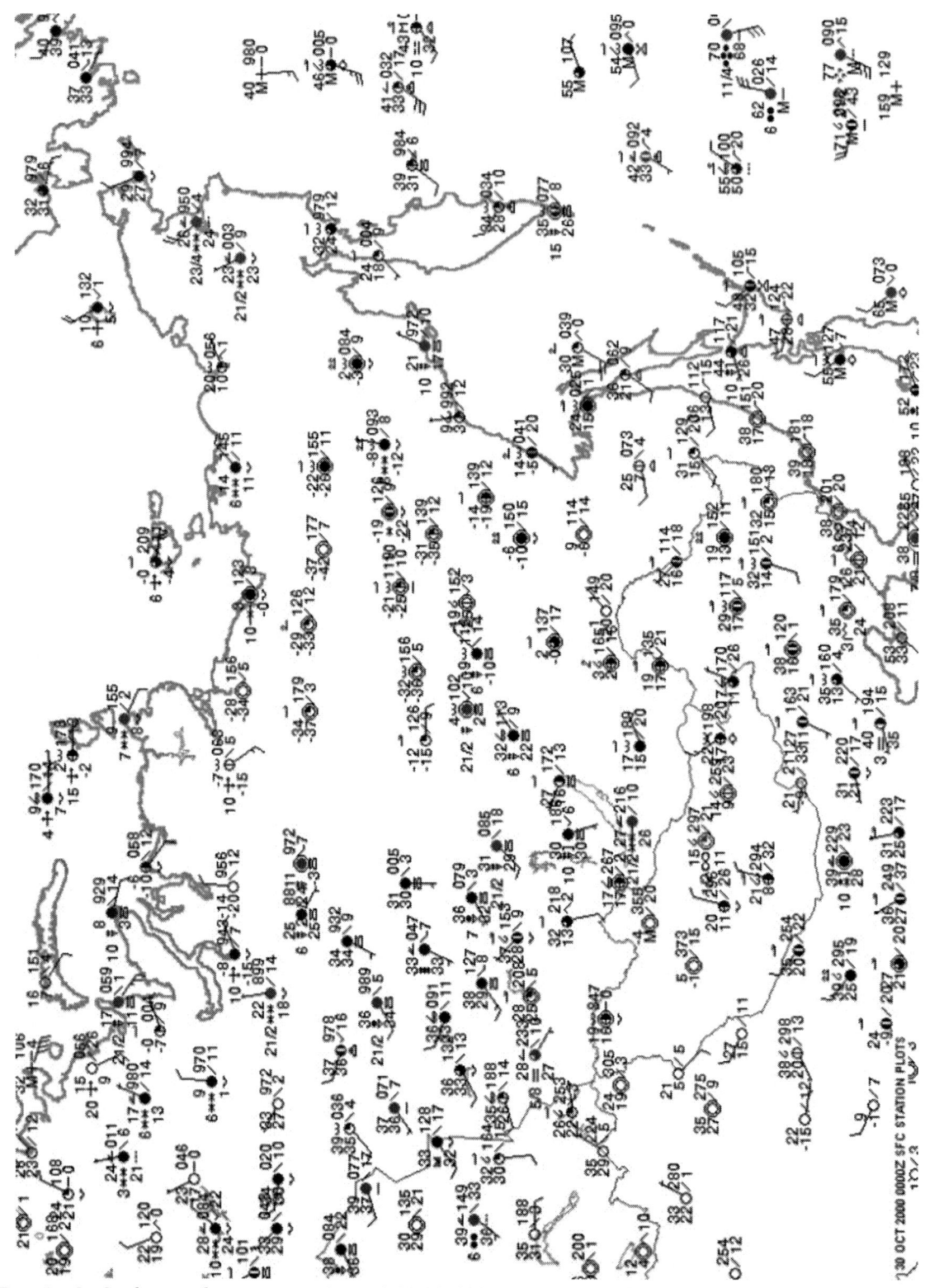

Exercise 3. Surface analysis exercise, Asia, 10/30/00, 0000 UTC (9 am China standard time).

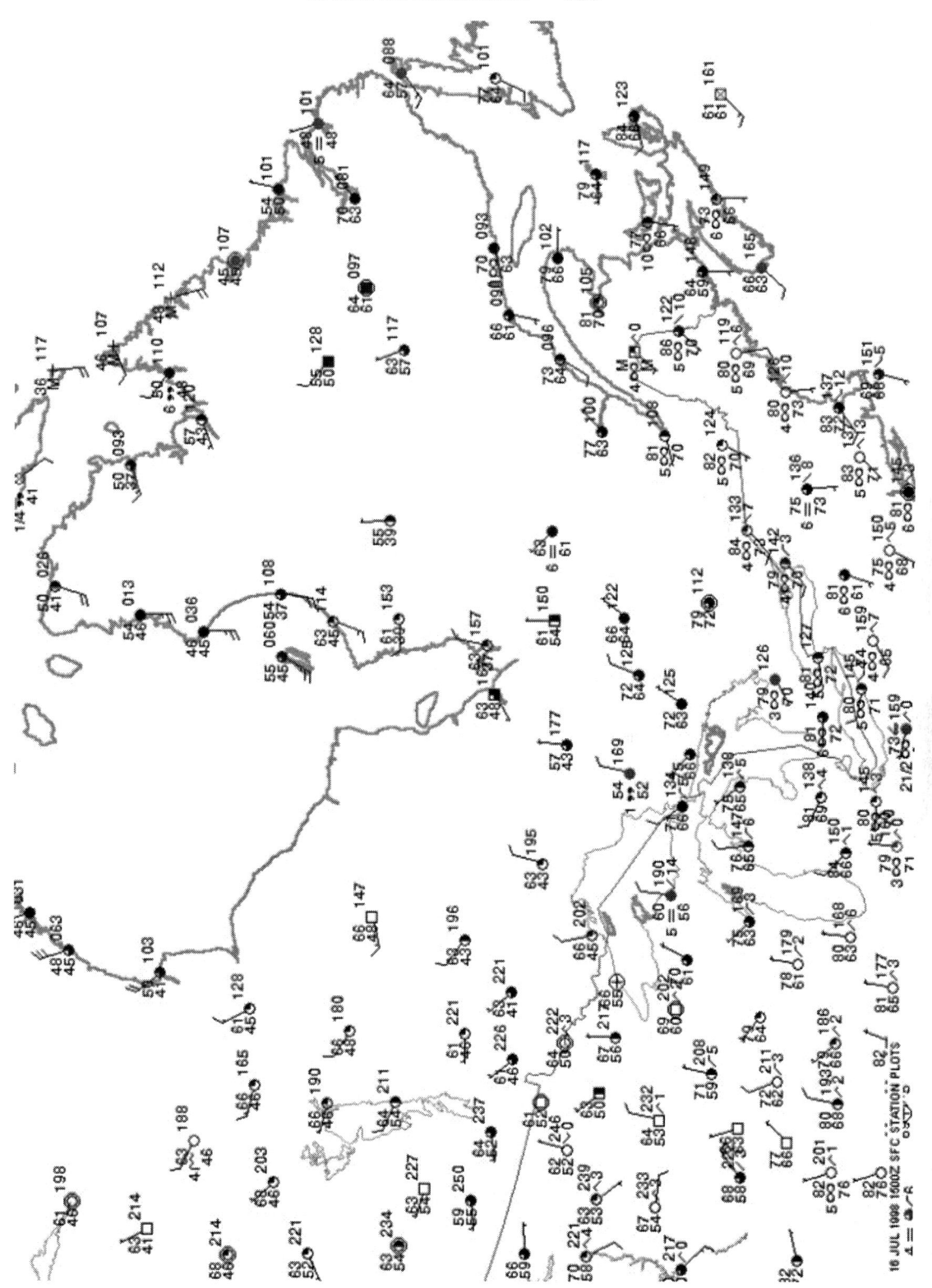

Exercise 4. Surface analysis exercise, eastern North America, 7/16/98, 1500 UTC (11 am EDT).

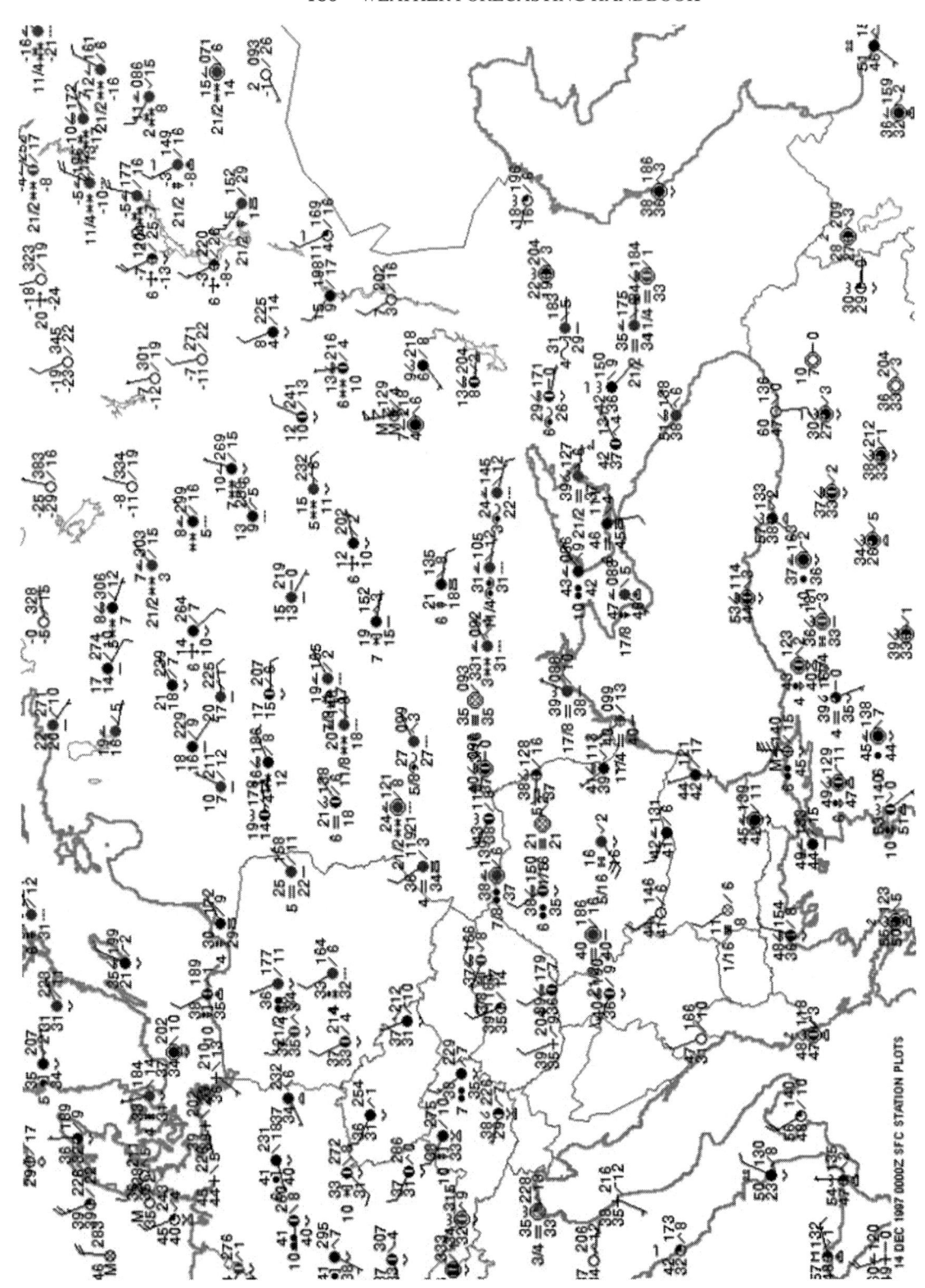

Exercise 5. Surface analysis exercise, western Eurasia, 12/14/97, 0000 UTC (3 am Moscow time).

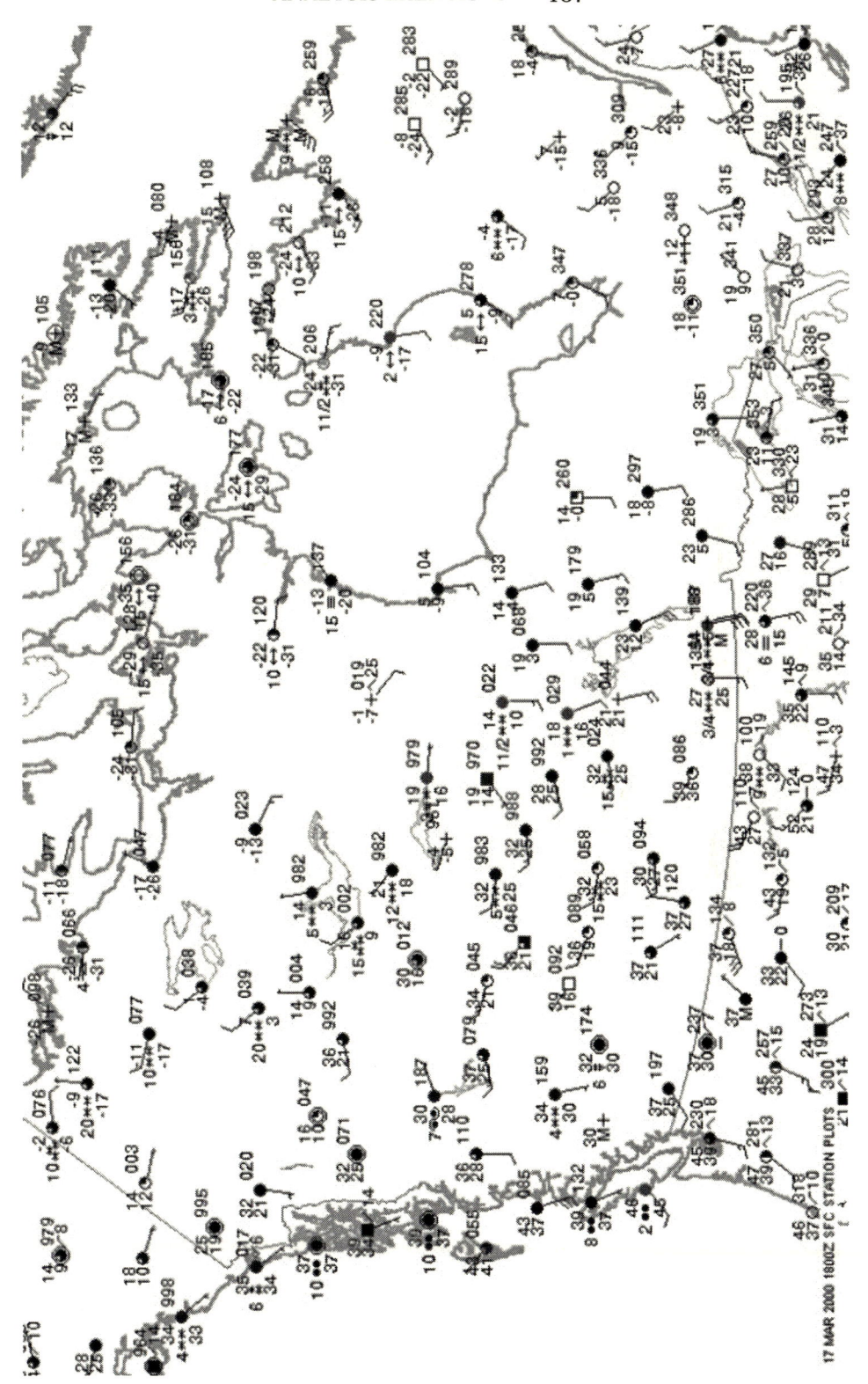

ANALYSIS EXERCISES • 187

Exercise 6. Surface analysis exercise, western Canada, 3/17/00, 1800 UTC (11 am MST).

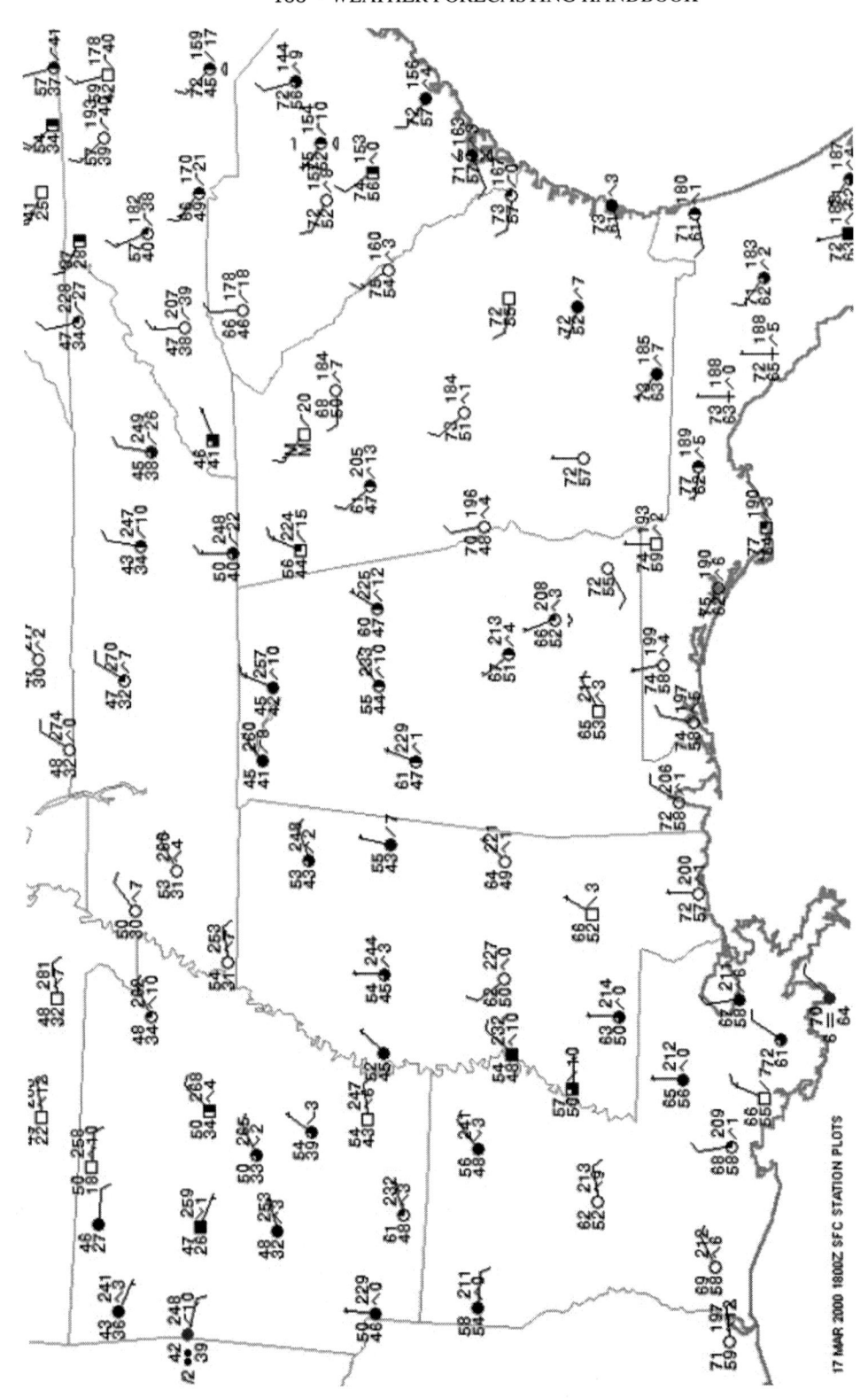

Exercise 7. Surface analysis exercise, southeastern U.S., 3/17/00, 1800 UTC (1 pm EST).

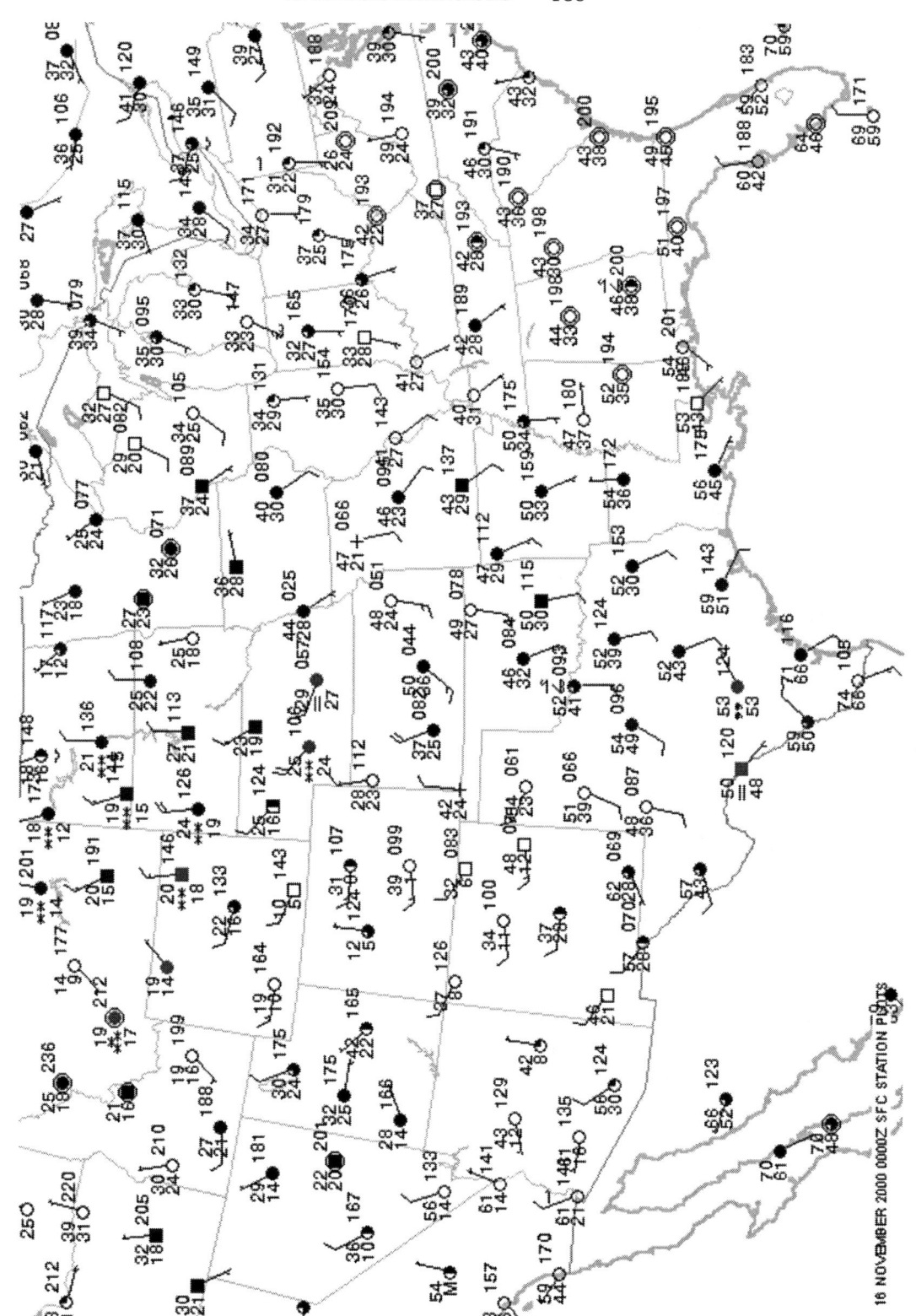

Exercise 8. Surface analysis exercise, U.S., 11/16/00, 0000 UTC (7 pm EST). Also see 500 mb chart for this same chart time on Exercise 10 and 300 mb chart on Exercise 11.

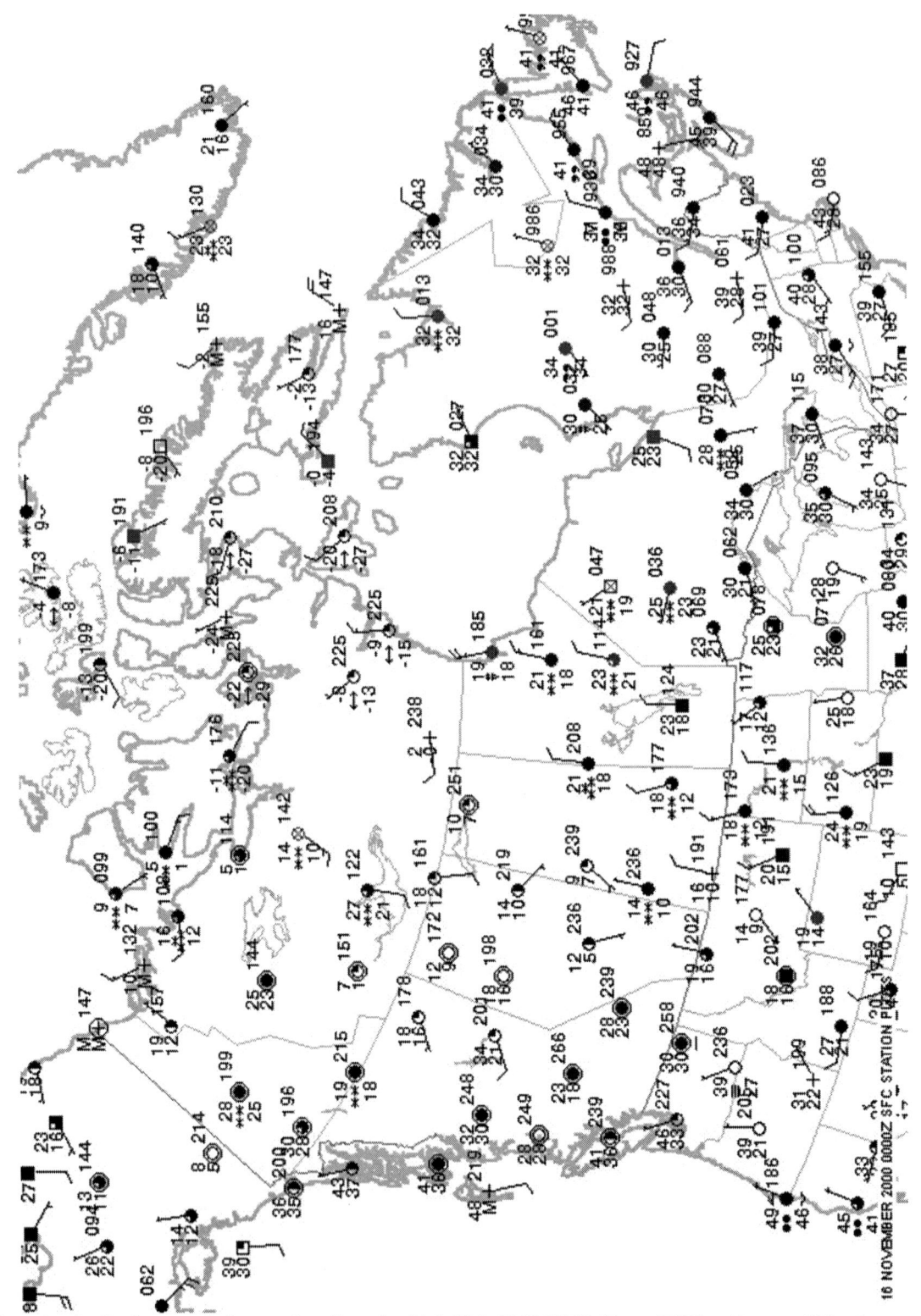

Exercise 9. Surface analysis exercise, Canada, 11/16/00, 0000 UTC (7 pm EST). Also see 500 mb chart for this same chart time on Exercise 10 and 300 mb chart on Exercise 11.

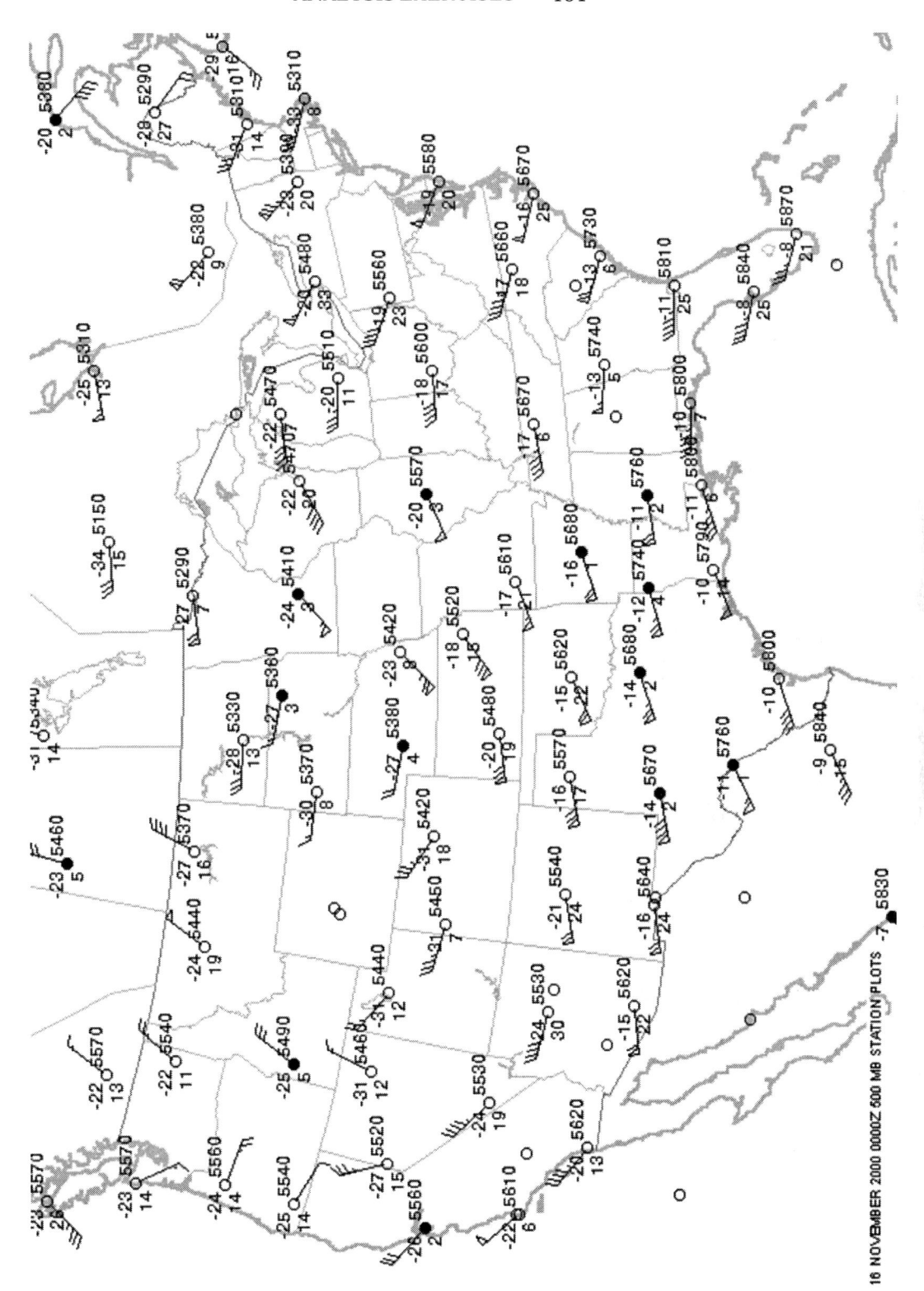

Exercise 10. 500 mb analysis exercise, North America, 11/16/00, 0000 UTC (7 pm EST). Also see surface charts for this same chart time on Exercise 8 and 9 and 300 mb chart on Exercise 11.

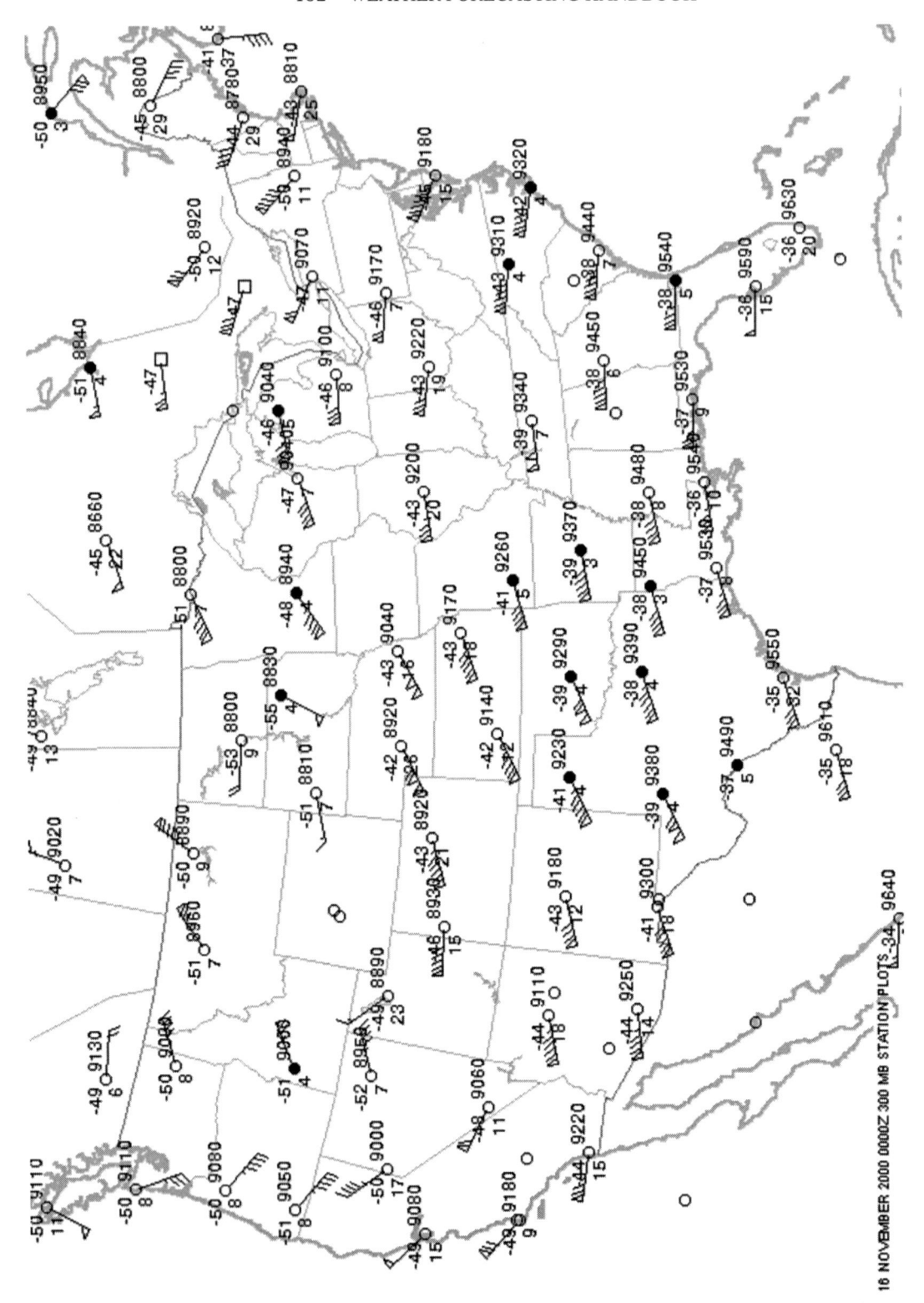

Exercise 11. 300 mb analysis exercise, North America, 11/16/00, 0000 UTC (7 pm EST). Also see surface charts for this same chart time on Exercise 8 and 9 and 500 mb chart on Exercise 10.

ANALYSIS EXERCISES • 193

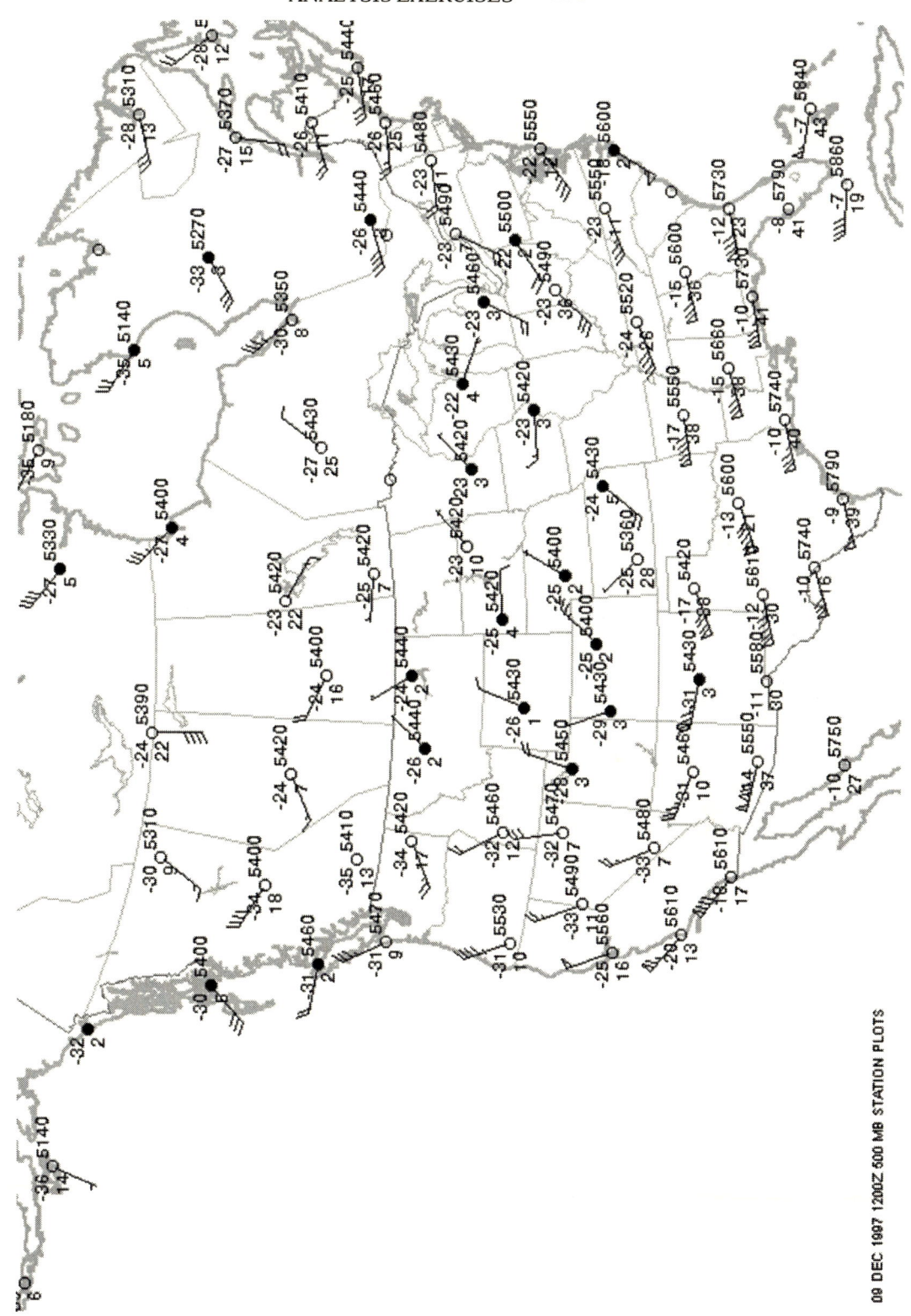

Exercise 12. 500 mb analysis exercise, North America, 12/9/97, 1200 UTC (7 am EST).

Index

Symbols

100 mb 11
1000 mb 10
200 mb 11
20R85 rule 133
300 mb 11
30R75 rule 133
500 mb 11
700 mb 11
850 mb 11

A

absolute instability 63
absolute vorticity 98
adiabat
   dry 49
   wet 49
adiabatic 59
advection
   thermal 95
advection lobe 102
ALSTG 22
altimeter setting 22
altocumulus 26
altostratus 25
anafront. *See* front: active
analysis
   numerical 161
angular velocity 64
anvil 25
AVN. *See* Spectral model

B

baroclinic
   high 122
   low 121
baroclinic instability 121
barotropic
   high, cold-core 112
   high, warm-core 113
   low, cold-core 111
   low, warm-core 112
BKN 23
bomb 122
boundary errors 167
boundary layer 67
bow echo 130
BRN 137
BRN shear 137
broken 23
BWER 43

C

CAA. *See* advection: thermal
CAPE 136
ceiling 23
centrifugal force 65
Changma 153
chinook 60
CIN 137
CINH 137
cirrocumulus 26
cirrostratus 26
cirrus 26
classic supercell 131
clear 23
cloud
   amount 23
   height 23, 26
   total amount 23
   type 24
CLR 23
cold air advection 96
cold dome 92
cold front 75. *See* front: cold
cold occlusion 77. *See* occlusion: cold
cold sector 74
combination models 162
condensation 59
conditional instability 63
conditional symmetric instability 106
confluence 90
contour gradient 65
contour gradient force 65
contours 11
convective feedback 167
convergence 90
Coriolis force 11, 64
couplets 43
cross totals index 139
CSI 106
CT 139
cumulonimbus 25
cumulus 24
cutoff high 113
cutoff low 111
cyclostrophic wind 66

D

deep easterlies 152
density 5
deposition 59
derecho 130
deviant motion 133
dewpoint 22
dewpoint depression 6
dewpoint temperature 6
difluence 89
divergence 88
domain
   global 163
   limited 163
double eyewall 157
dry adiabat 49
dry adiabatic lapse rate 60
dryline 78
   location 79
   movement 79
   structure 78
dust 79
dynamical models 162

E

easterly waves 153
ECMWF model 166
EHI 137
elevated mixed layer (EML) 78

Elvis' UFO 10
EML 78. *See* elevated mixed layer (EML)
energy-helicity index 137
equation of state 7
equatorial trough 151
equatorial wave 151
equivalent potential temperature 106
eta 164
evaporation 59
extratropical cyclone 121
eye 157
eyewall 157

F

Feng Yun 46
ferrel cell 12
force 5
freezing 59
friction 66, 92
frictional force 65
front 72–74
  active 75
  anafront. *See* front: active
  inactive 75
  inversion 73
  katafront. *See* front: inactive
  location 72
  movement 74
  slope 74
  stationary. *See* front: quasistationary
  surface 73
frontal inversion 73
frontal low 121
frontogenesis 74
frontolysis 74

G

geostrophic wind 65
GMS 46
GOES 45
GOMS 46

gradient wind 66

H

Hadley cell 152
hadley cell 11
heat low 112
Henry's rule 111
hodograph 45, 47, 49
hodographs 49
HP supercell 132
hurricane 112, 155
hydrostatic equation 8, 60
hypsometric equation 8

I

ID method 133
inflow 134
infrared imagery 46
initialization 161
INSAT 46
instability 127
intertropical convergence zone 151
isentropic analysis 103
isentropic lift 105
isentropic surfaces 103
isothermal vorticity advection 100
ITCZ 151

J

jet 80
jet streak 93
jet stream. *See* polar front jet

K

K-Index 138
katafront. *See* front: inactive
KI 138
kinetic energy 95, 121

L

land breeze 80
lapse rate 62

dry adiabatic 60
wet adiabatic 60
latent heat 60, 112
latitude 64
level of non-divergence 92
LFM 163
LI 138
lift 127
Lifted Index 138
lightning
  detection of 54
  positive strikes 54
LLJ 82. *See* low level jet
long wave 85–88
  characteristics 86
  number 86
  polar vortex. *See* polar vortex
  ridge 87
  scale 86
  trough 88
low level jet 82
LP supercell 131

M

MCC 130
MCS 131
melting 59
mesoscale 9
mesoscale convective complex 130
mesoscale convective system 131
mesosphere 9
METAR 21
METEOSAT 45
microscale 9
mid-tropospheric cyclones 154
mixing ratio 6
  and SKEW-T's 49
moisture 96, 127
momentum 79, 106
monsoonal trough 151
multicell
  cluster 129

INDEX • 197

line 129
MCC. *See* MCC

N

National Center for Environmental Prediction 161
NCEP 161
negative vorticity advection 99
NEXRAD 41, 43
NGM 164
nimbostratus 25
NIVA 100
non-divergence
   level of. *See* level of non-divergence
NVA 99. *See* vorticity: advection

O

obstructions to vision 23
occluded front. *See* front: occluded
occlusion 111
omega equation 99
outflow cirrus 158
OVC 23
overcast 23

P

parcel 59
PFJ 80. *See* polar front jet
phase changes 59
   condensation 59
   deposition 59
   evaporation 59
   freezing 59
   melting 59
   sublimation 59
plateau high 113
polar cell 12
polar front jet 80
polar high 113
polar orbiters 46
positive CG lightning 54

positive vorticity advection 99
potential energy 95, 121
potential temperature 7
pressure 4, 22
Pressure coordinate system 10
pressure falls 91
pressure gradient 65
pressure gradient force 65
pressure surface 10
prevailing visibility 23
pseudo-adiabatic 60
PVA 99

Q

Q-vector 102
QFE 22
QLM. *See* Quasi-Lagrangian model
QNH 22
quadrants
   divergent/convergent 94
   left front 95
   left rear 94
   right front 95
   right rear 94
Quasi-Lagrangian model 165
quasistationary front 77. *See* fronts: quasistationary

R

radar 41
RAFS. *See* NGM
reflectivity 41
relative humidity 7
relative vorticity 98
rising motion 91

S

Saffir-Simpson 157
SAO 21
satellite
   infrared imagery 46
   visible imagery 46
   water vapor imagery 47

saturation mixing ratio 7
scale 8
   mesoscale 9
   microscale 9
   planetary 8
scattered 23
SCT 23
scud 24
sea breeze 80
Sea-level pressure 22
sectors 74
self-development 121
severe thunderstorms 41
SFD. *See* forecast discussion
shallow easterlies 152
shear lobe 101
short wave 88
   origin 88
   scale 88
Showalter Index 138
SI 138
sinking motion 92
sloshing 79
SLP 22
Spectral model 165
spectral models 163
speed convergence 90
speed divergence 89
spiral bands 157
split 130
squall line 128
SRH 137
stability 60–64, 62
station pressure 22
stationary front. *See* fronts: quasistationary
statistical models 162
STJ 82. *See* subtropical jet
storm movement 134
storm-relative helicity 137
stratocumulus 24
stratosphere 9
stratus 24
subgeostrophic wind 66, 93
sublimation 59
subsidence 62

subtropical high 113
subtropical jet 82
subtropical ridge 152
subtropical stationary fronts 152
supercell 129
  classic 131
  HP 132
  LP 131
supergeostrophic wind 66, 94
surface aviation observation 21
SWEAT 138
SYNOP 21
synoptic observations 21
synoptic scale 8

T

temperature 5, 22
terrain 92
thermal advection 95
thermal low 112
thermosphere 10
thickness 96
thunderstorms
  severe 41
tornadoes 158
total totals index 138
trade wind inversion 152
trade wind trough 151
trade winds 152
transverse circulations
  of jet 95
tropical cyclone 112, 154
tropical upper tropospheric
  troughs 154
troposphere 9
trough 77, 79
TT 138
TUTT 154
TUTT low 154
typhoon 112, 155

U

UKMET model 166

unicell 128

V

VAD 43
velocity 43
ventilation 134
vertical motion 60, 90
  rising. *See* rising motion
  sinking. *See* sinking motion
vertical totals index 138
virtual temperature 7, 62, 96
visibility 23
visible imagery 46
vort lobe 102
vorticity 88, 96
  advection 97
  advection lobe 102
  and extratropical cyclones 121
  and jet streaks 100
  components 97
  isothermal advection 100
  lobe 102
  negative 97
  shear lobe 101
  vertical motion 99
VT 138

W

WAA. *See* advection: thermal
warm air advection 96
warm front 75. *See* front: warm
warm occlusion 77. *See* occlusion: warm
warm sector 74
warm sink 91
water vapor imagery 47
WBZ 139
weather 23
WER 43
wet adiabat 49
wet adiabatic lapse rate 60
wet bulb temperature 6
wet-bulb thermometer 22

wet-bulb zero 139
wind 22, 65–68
  cyclostrophic 66
  geostrophic 65
  gradient 66
  subgeostrophic 66
  supergeostrophic 66
wind vectors 133
WSR-57 41
WSR-74 41